INTEGRATED ENVIRONMENTAL PLANNING

To those that "sat the bench" and still managed to get into the game

Integrated Environmental Planning

James K. Lein

Department of Geography

Ohio University

Blackwell
Science

© 2003 by Blackwell Science Ltd
a Blackwell Publishing company

Editorial Offices:
Osney Mead, Oxford OX2 0EL, UK
    Tel: +44 (0)1865 206206
Blackwell Science, Inc., 350 Main Street, Malden, MA 02148-5018, USA
    Tel: +1 781 388 8250
Blackwell Science Asia Pty, 54 University Street, Carlton, Victoria 3053, Australia
    Tel: +61 (0)3 9347 0300
Blackwell Wissenschafts Verlag, Kurfürstendamm 57, 10707 Berlin, Germany
    Tel: +49 (0)30 32 79 060

First Published 2003 by Blackwell Science Ltd

Library of Congress Cataloging-in-Publication Data

Lein, James K.
    Integrated environmental planning / James K. Lein.
            p. cm.
    Includes bibliographical references and index.
    ISBN 0-632-04346-6 (pb.: alk. paper)
        1. Environmental management—Planning.   I. Title.

    GE300.L453 2002
    363.7'05—dc21

                                        2001043951

ISBN 0-632-04346-6

A catalogue record for this title is available from the British Library.

Set in 9.5/12 pt Palatino
by SNP Best-set Typesetter Ltd., Hong Kong
Printed and bound in Great Britain
by TJ International, Padstow

For further information on
Blackwell Science, visit our website:
www.blackwell-science.com

Contents

Preface

Many fields contribute to the study and practice of environmental planning and the contents of courses on the subject typically reflect this diversity. While some may view this as a type of bias where one instructor's background and training suggests the only path the topic may follow, diversity suggests richness and reflects the truly multidisciplinary nature of environmental planning and problem-solving. I do not believe that there is a single best approach to this subject – the topic is as dynamic as the landscape on which planning is practiced. Although environmental planning enjoys a 30-year history as a formal course of study, it remains an evolving subject that changes as our awareness and understanding of the environment grows. I do not pretend to have all the answers in this book, nor do I wish to provide the final word on the topic. Rather, I set out to write this book with the ambitious goals of placing environmental planning into a more contemporary context and differentiating the environmental approach to planning from subjects such as resource management, environmental engineering, and landscape analysis. In this book, my principal aim was to bring together a selection of pertinent themes and topics that support my approach to environmental planning. One of the greatest challenges, I soon discovered, was the enormity of the task of synthesizing and integrating a disparate collection of material that forms the foundation to the subject. Out of necessity, I became very selective, identifying the approach I considered important to the student, then including only that material I considered germane. In reading this text others may find that I have devoted too much discussion to one subject and not enough to another. Fortunately, the field of environmental planning is large enough to support others' contributions, and I hope this book will sustain interest in environmental planning and encourage others to produce treatments on this topic as well.

I look upon this text as an introduction to the field that links the process of planning to the landscape and presents the "need to know" information that will ensure that environmental concerns remain part of the planning process. I have also connected planning to the technologies that support the analytic and integrative aspects of the subject. Through the inclusion of GIS and computer modeling, it is my aim to close the gap between these technologies and their application in order to help the student recognize that these methods have merit and that they complement the activities of the environmental planner. The audience of this book can be as varied as the subject itself. While this text was written primarily to reach undergraduates majoring in geography, environmental studies, and regional planning, I hope that the material will also be of interest to graduate students specializing in environmental planning and management, and environmental professionals as well.

Finally, while I may have been the person that put these words onto the page, there were many others who helped me along in this process. First, I need to recognize the support and patience of my family – my wife Christine, and my daughters Adriana and Elena: none of this would have been possible without you, and I would never have embarked upon this project had it not been for your motivating influences. Next, I wish to thank those that reviewed earlier drafts of this work. Your comments were helpful and I know this book is better because of them. Finally, I'd like to thank that little voice in the back of my head that kept telling me to stick with it even though there were times when chucking it all and selling T-shirts by the beach seemed like a better prospect.

James Lein
Athens, Ohio

CHAPTER 1

The Nature of Planning

Each of us engages in the process of planning, whether it is in the context of our next vacation, our eventual retirement, or the selection of courses for our next academic term. Planning has been observed to be a fundamental human activity, and suggests an important strategy that helps us consider possible outcomes before we commit to a specific course of action (Catanese & Snyder, 1988). Yet the concept of planning and the intellectual tools that we apply in our daily lives are substantively different from how the environmental professional practices planning. When applied to the environment, planning is concerned with the problem of reconciling environmental functioning to broadly defined stakeholders, each with diverse and often conflicting interests. The goals of planning when placed into this arena and the means to achieve them can be highly uncertain, and the results of most environmental plans can be realized only after long periods of time have passed. In this chapter we will set out on our exploration of environmental planning by looking first at the intellectual tools that drive it. From here we can examine what it means to plan and how we can organize our thinking to focus our recommendations about the future in the way that is most productive. Throughout this introduction to the nature of planning, emphasis is given to the idea that planning is problem-driven, information dependent, and never an absolute or perfect answer.

Common themes and common problems

To begin our exploration of environmental planning we can examine a few hypothetical situations that help place the problem of planning and its role in society in its proper context. Above all, these examples remind us that the environment in which planning takes place is complicated by the numerous voices, issues, and opinions that must be addressed as alternatives and recommendations for the future are expressed.

If you build it, they will drive, then what?

In this first example, a land developer approaches officials in a small community with a proposal to construct a regional mall. The community sees this proposal as an important opportunity to increase employment and to encourage economic development. The necessary permits are granted and construction begins. Once the project is completed and the mall opens, the single road leading to the new facility quickly becomes congested with traffic and the need to widen this major artery to accommodate the activity generated by the mall is quickly recognized. Widening the road, however would require businesses along one side of the street to lose valuable land that is presently used for parking and product display. On the opposite side of the street, road widening would mean that a row of 15 regionally significant trees would need to be removed. The trees are considered by many in the community to be aesthetically pleasing, and indicative of the rural small-town atmosphere that is a source of pride in the community.

In the absence of any road improvements, neighborhoods paralleling the main mall access road have noticed a steady increase in traffic as shoppers explore shortcuts to avoid traffic snarls and delays. The increased traffic elevates noise levels and introduces serious safety concerns for residents in the affected areas. Residents of these neighborhoods would like to see the mall access road improved to alleviate the traffic, noise, and safety problems. Businesses along the access street would also like a wider street, but not if it means losing front footage they need for off-street parking. Residents of the community who travel along the street leading to the mall would also like a wider road since it would reduce travel time and congestion, but many would hate to see the tree-lined street destroyed. A "save our local environment" group opposes any plan to widen the road on the grounds that it would encourage more traffic, and the loss of historically significant trees would degrade the sense of "place" in the community. They advocate greater funding for alternative forms of transportation and press civic leaders to consider more sustainable forms of land development.

As decision-makers weigh options for the future, if a mechanism had been in place that would have foretold the possibility of these consequences, a more appropriate decision could have been made that would have avoided some, if not all, of the problems that are now more difficult to resolve.

Homes on the range

Across the river, residents of a hillside housing development are concerned by a proposal to expand their subdivision. This neighborhood, comprised primarily of upper-middle income professionals, is isolated in a wooded area, quiet, surrounded by habitat native to the region. Residents in this area enjoy the proximity to open space and use this area as a local recreation resource. However, there is an intense demand for single-family housing in the region and the land area in question is extremely attractive, accessible, and well suited for the types and densities of development proposed. Construction of homes in this area would satisfy local demand, enhance the community tax base, and reduce land market pressures that have already begun to elevate rents and house prices in the region. Being somewhat isolated, the area is served only by a single two-lane road that traverses over a sequence of small ridge tops and passes by a series of houses dispersed in this historically rural area. Present residents are concerned that the addition of more homes would increase vehicle trips and increase traffic along this narrow corridor. This would lead to noise and safety problems and alter the atmosphere of this quiet rural setting. Others in the region are concerned over the loss of a valuable amenity resource that by itself provides important habitat functions for deer and other animals living in the region. Others are worried that new development cannot be accommodated by the existing water and sewage system and that inadequate water pressure may degrade the flow of fire hydrants in the area and present a very real risk to public safety. In the absence of a clear plan, the objectives and needs of the community cannot be adequately balanced by the equally important need to maintain the viability of the larger environmental system.

Down by the sea

Finally, there is the example of the community located in a coastal area popularized by summer vacation cottages, and where recreation and tourism have traditionally supported the local economy. To maintain the rural setting and atmosphere, building densities were kept purposely low and the dispersed pattern of settlement required most homes to use septic systems to treat domestic waste water. Over time, growth pressures coupled with urban encroachment emanating from the large urban center only 40 miles away have transformed this community to an ex-urban settlement well within the commuter shed of the rapidly growing metropolitan region to the north. Population now resides here year round, and because of the widespread use of septic systems, coupled with the community's location on predominantly sandy soils with a water table close to the surface, serious water quality problems have become recognized. Cultural eutrophication and overloading is common in the marshlands and estuaries along the coast and this has adversely impacted wetland functioning and altered habitat. Residents now complain of the bad smell, algae growth, and other nuisances, while environmental groups ask for growth controls and demand steps be taken to mitigate the adverse impact development is having on the coastal region.

Each of these illustrative vignettes introduces several common themes that focus our attention on the nature of planning and problems that encourage an environmental approach. This environmental response to planning recognizes the need to achieve a balance between human requirements to exploit the landscape to satisfy societal wants, with the equally important need to maintain and enhance environmental quality. Thus, unlike community planning, urban planning, or its variants, environmental planning is uniquely concerned with understanding the connection between human landscape and the ecological and physical processes that directly and indirectly sustain our existence. By employing this understanding in the design of plans and policies, better plans can be developed and more sustainable human patterns can be crafted. A set of common ideals or themes helps project that understanding.

The first major theme underscored in the vignettes introduced above is that of change. As a theme, change reminds us that the world we inhabit is dynamic and that we (humans) are always responding to or encouraging change in our world. This change may be purposeful or inadvertent; nonetheless it is a process that we live within, and a process that directs us as we attempt to direct it. The second theme is that of consequence. Consequence describes the culmination of a sequence of events that represent a pattern or reality that we must confront. It points to the fact that processes produce events and events take on a form and become real. The third theme is uncertainty. As events and decisions reshape our world,

we recognize that change is underscored by an element of uncertainty. In this context uncertainty points to the unknown and often the unknowable. It also suggests to us that change, the processes it defines, and the events that materialize are always subject to our ignorance. The next theme is choice. Although the presence of choice may not always be obvious, those consequences result from a given alternative: a possible arrangement of things that contributed to the events which ended with the reality we see. Of course we may not always know that ahead of time, so we often wait for events to unfold. What's more, we may be uncertain as to which alternative is the best choice and how events may be connected as they drive us toward change. However, in most cases no one wants to wait for the end to be realized. We'd all like to know in advance what the possible outcomes might be, or at least be given a hint as to what we might expect. This universal human quality introduces the final theme that motivates planning: risk. Risk explains the possibility of being wrong and what being wrong may mean in human and environmental terms. Although risk can be defined more precisely later in this text, for the moment we can think of risk as the proverbial fork in the road where we have to decide which path to take, and learn to understand the implications of the wrong choice and accept the poor alternative and the adverse consequence that may follow.

In each of the illustrative examples presented above human beings introduced or suggested a change. From that change an outcome was produced, an alternative was selected, an element of risk could be identified, contrasting perspectives were shown, and each ended with an uncertain reality. Those realities invite more opportunities to affect change, consider a set of alternative actions, deliberate over uncertainty, and make judgments about risk. For most of us, none of the outcomes presented in our illustrative scenarios are desirable. In the first example we are left with a street that cannot be widened since it will compromise local businesses on one side, and a line of historically significant trees on the other. In the meantime, neighborhoods suffer the consequence of increased traffic and possible risks to safety.

Wouldn't it have been easier to have "seen" these consequences well before we introduced this change, looked critically at all the factors involved, and considered the implications of choice?

In the second example we are asked to decide between the need for housing for people and the loss of habitat for deer. In addition we are required to consider the value of land for its aesthetic use and realize its value for that purpose, while also evaluating the importance of meeting basic human needs. In this hypothetical example, we'd like to know what the alternatives are, how well they meet the human need for affordable housing while also protecting important habitat qualities for functioning, sustainable environment. Can we acquire that information to guide us to make the appropriate choice?

Again, in our final example we are dealing with change and the external forces that introduce effects that we don't always see. Here, we are asked to examine the connections between human-driven processes of change and how they interact with those of the environment. We are also required to look at the cumulative effects of change and how they create a reality that would be difficult to see one person and one transformation at a time. Above all, this example, as do the others, suggests the importance of seeing the future or at least directing the future toward a compatible state. It is this future orientation and the need to be proactive that embodies the concept of planning.

The concept of planning

The concept of planning is difficult to define in precise terms. Perhaps at its most fundamental level planning can be described as a universal skill that involves the consideration of outcomes before a choice is made among alternatives (Feldt, 1988). To illustrate this idea, consider the desire of Anytown, USA, to preserve open space for recreational uses. Open space and recreation are fairly well understood ideas, but the decision of which lands to preserve as open and for what recreational uses is not a simple matter. Should lands be pre-

served for hiking or off-road vehicle use? How do we decide? What if we are wrong? Therefore, we can refine our definition to describe planning as a method for reconciling choice under conditions of risk and uncertainty. However, regardless of definition, a central element of planning is the desire to direct change in order to produce a beneficial consequence at some point in the future. In this sense we can think of planning as a "vision." For example, in a small town faced with the pressure to grow and develop, planning becomes the means by which this community sees itself and expresses how it wants to appear in the future. This future orientation is a critical aspect of what planning means, although it can be a perspective that is easily forgotten when reactive thinking dominates public-policy making agendas. Because planning takes place along a time continuum, it is more than simply a skill, it is a future-oriented activity that contains its own unique formalisms and directives.

From this simple definition, the scope of what planning is and what it means begins to take shape. Here we may introduce several pragmatic considerations to extend our definition. First is the realization that planning as an activity is motivated by specific goals and objectives. This idea suggests that when one undertakes the intent to plan, one necessarily defines a course of action. As such the plan describes a type of decision-making where goals and objectives are used to help select among alternative solutions. The implication, based on the above, is that planning is purposeful and defines a continuing process that helps organize our thinking. Above all, as Barlowe (1972) reminds us, planning is the opposite of improvising. Thus we can consider the activity of planning as a form of proactive decision-making where the risks and uncertainties of the future are minimized and a course of action or program takes form that facilitates the wise allocation of important and potentially scarce resources. Recent events surrounding the California energy "crisis" illustrate how unanticipated events and the cumulative effects of "poor" planning can leave few options and force policy-makers into a reactive posture. By minimizing risk and uncertainty, planning supports the belief that the future can be

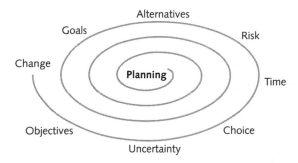

Fig. 1.1 The features of the planning problem.

controlled, albeit in a limited and selective way, so that societies' "vision" can be realized (Fig. 1.1).

As Levy (1997) asserts, the need for planning in this context simplifies to two basic words, interconnectedness and complexity. If the earth were sparsely populated and the technologies we lived by simple, there would be little need for planning (Levy, 1997). The observed fact, however, is that planet earth is not sparsely populated and our technologies are sufficiently complex that without consideration for the future and the wise management of resources an orderly progression of society cannot be assumed. A reasonable person may argue that as society commits itself to a more intensive use of the earth's surface, planning becomes a necessity. However, there are differential levels and degrees to which the concept of planning is applied and practiced which contributes to many of the problems society confronts. Therefore, through planning we can prioritize the goals and needs of a community and manage scarcity with improved efficiency. For that reason it becomes convenient to conceptualize the question of why we plan as the process which evolves to guide us toward the future.

The process of planning

Planning is both a logical process and a methodology that defines a series of components that direct our attention toward four interrelated activities:

1 Establishment of goals and objectives.
2 Collection and analysis of information.

3 Evaluation of alternative courses of action.
4 Recommendation of a course of action.

A general planning procedure is typically comprised of a number of stages or phases executed systematically over a specific time schedule. Although there is a tendency to conceptualize the activity of planning as a clearly defined linear sequence, in actuality the stages are not always followed in a rigid sequential fashion – and nor should they always be. Rather the process may evolve iteratively with considerable elaboration and refinement along the pathway to the solution. One convenient way to examine the process of planning is through the lens of rationality. The fundamental stages of the rational approach to planning include:

1 Identification of the problem and determination of need.
2 Collection and analysis of data.
3 Development of goal and objectives.
4 Classification and diagnosis of the problem and surrounding issues.
5 Identification of alternative solutions.
6 Analysis of alternatives.
7 Evaluation and recommendation of actions.
8 Development of an implementation program.
9 Surveillance, monitoring and evaluation of the outcome.

This nine-step process is illustrated in Fig. 1.2 and presents the logical flow of tasks together with several non-linear elements that suggest places along the sequence of phases where review and refinement may be encouraged. According to this rational approach to planning, each phase in the process consists of numerous substeps that vary in detail in relation to the nature of the problem.

Real-life planning decisions do not always follow the rational approach (Leung, 1989). Several reasons can be offered to explain why. First is the realization that many planning decisions are reactive in nature and have a much shorter time horizon and scale than long-range planning enjoys. Secondly, there is often a lack of resources that frustrates attempts to create carefully articulated, systematic methodologies. Lastly, the structure of the rational approach may not fit with the nature of the planning problem under consideration. All

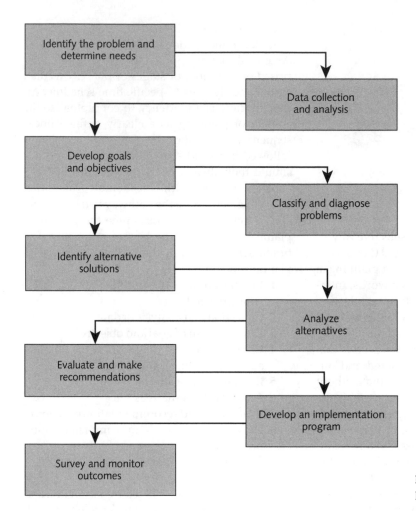

Fig. 1.2 The general process of planning.

too frequently, the complexity, interconnectedness, and uncertainty that surround planning contribute to the development of problems that are poorly structured or "wicked." Such problems often keep transforming themselves or contain elements that are not well understood. In these situations the qualities of the problem suggest that a rigid adherence to a particular mode of thinking or analysis may be unrealistic and undesirable.

However, when viewed as a process, planning, while methodical, must remain sensitive to changing needs and circumstances as dictated by the problem (Leung, 1989). In this context, regardless of complexity, the availability of resources, and time pressures, planning always involves the careful definition of the problem, the develop-

ment of goals and objectives, the identification and analysis of alternatives, the collection of data, and the implementation of a program or course of action. By connecting these analytic components of the planning process to the substantive issues that motivate us to take action, an outline to direct planning can be developed. That procedural outline and its salient characteristics are examined below.

Problem definition and expressing needs

Formulation of the right planning problem is the pivotal beginning place in the process of planning

(George, 1994). Problem formulation begins with the awareness of need, where need may be expressed in very specific terms, such as the need to widen a road to accommodate increased traffic, or can be articulated in a much broader sense, such as a community's need to enhance environmental quality or preserve open space. Needs, however, can be very elusive and difficult to express, particularly in situations where community aspirations conflict. Therefore, identifying the problem may mean more than simply problem-solving, but also problem avoidance. In practice, planning is most frequently used to address perceived problems and to ensure that these problems do not occur in the future (Leung, 1989). However, the question to consider when defining the problem involves careful consideration of the distinction between a problem and the "right" problem. This is not unlike the situation encountered in areas such as northeastern Ohio that are experiencing the loss of agricultural land. It's not that farms are disappearing, it's that annexation and land-tax policies encourage homes to appear instead.

While it may seem an oversimplification, a problem when viewed through the lens of planning defines the difference between expectation (farmland) and reality (suburbanization), and those expectations and the reality perceived are often heavily value laden. Consequently, problems tend to be identified by reference to an expectation or goal. Expectations may be as basic as a road that is free of traffic congestion, a development proposal that will not adversely affect wetland functioning, or a housing stock that is affordable to middle income groups. Expressions of a problem may be based entirely on a description of symptoms, the outgrowth of previous studies that have discovered an issue that requires action, or an idealistic affirmation of community values voiced under specific conditions or circumstances. In each of these instance we must ascertain whether the problem is the right problem.

George (1994) notes that problem-solving is a ubiquitous human endeavor found in different facets of everyday life as well as in crisis situations. Planning is also a vital problem-solving endeavor; unfortunately, planning problems are not always solved successfully. While there may be numerous reasons to explain planning failures, solving the wrong problem is a surprisingly common factor (George, 1994). Solving the wrong problem is similar in concept to a Type III error in statistical hypothesis testing. In statistics, a Type III error explains the situation where the hypothesis tested has little relevance to the phenomenon under investigation. Therefore, when attempting to determine whether the right problem has been identified, it becomes important to recognize the fact that problems, in the abstract, are not real entities, but mental constructs. They explain or represent an unsatisfactory reality that is subjective and does not exist outside the perceptions and conceptualizations of the individuals confronting them (Smith, 1989). This point noted, problems are, in essence, the products of thought acting on environments that characterize elements of problematic situations that have been abstracted by analysis. The fact that a hillside poses a landslide threat is only problematic if we wish to subject that hillside to some form of human use. Likewise, an earthquake fault trending under a valley presents a problem only if its presence is unsatisfactory to the goal of developing that valley for high-density urban uses. Thus, before continuing along this line of reasoning, a distinction must be drawn between the problem, explained as a mental construct, and a problematic situation, which is an external reality. With respect to the planning process, problematic situations are the targets of identification, and from those situations problems are defined. Problematic situations can be arranged along a continuum that can be used to show how they may be approached and which solution strategies are most appropriate when considering their influence (Fig. 1.3). At the low end of this continuum are problematic situations that are puzzle-like and comparatively well defined (the need for additional parking spaces in a downtown shopping district). When confronted with problem situations of this type, goals can be clearly prescribed and solution strategies can be derived that will yield satisfactory results. At the opposite extreme lie the "messy" situations that are characterized by highly interrelated problems that interact with one another and tend to be difficult to decompose into more tractable descriptions (locating a site for

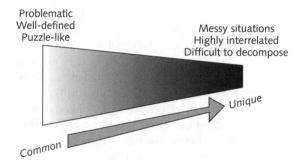

Fig. 1.3 The problem continuum.

the perpetual storage of low-level radioactive waste). Messy situations have numerous goals that often conflict, and this often makes it impossible to determine whether a satisfactory solution has been found. Between these two extremes lie problematic situations that are commonly referred to as "wicked" (Churchman, 1967). Although such problematic situations may not involve conflicting goals, they typically cannot be conceptualized in a unique fashion, and therefore tend not to have well-specified solutions.

The importance of drawing these distinctions when examining the nature of problematic situations arises from the fact that the more complex a problem situation is, the more ways it can be represented. To the planner, this suggests that problem definition involves both perception (seeing) and conceptualization (thinking). Therefore, how a problem is perceived and conceptualized will influence how it is represented; yet the strategies used to address the many possible expressions a problem may take may not address the "real" situation, since a representation can be incorrect, incomplete, or inappropriate. Put simply, an incorrect representation of a problem does not recognize any of the elements that constitute the problematic situation; an incomplete representation misses several elements; and an inappropriate representation ignores features salient to those affected by the situations (George, 1994). For example, in a plan to encourage the reintroduction of native plants into an urbanizing watershed, critical soil or microclimatic variables may have been omitted from the "model." Consequently, because important elements of the environment were left

out, the description of the habitat will not adequately reflect the real situation.

Successful problem definition rests with how problems are formulated. Through careful problem formulation the attempt is made to:

1 Conceptualize the problematic situation – forming an "image" of what is involved.
2 Arrive at a representation of the problematic situation – trying to explain what that "image" looks like.
3 Form a basis for generating solutions – looking for all the possible alternatives that might address the problem.
4 Develop a means to evaluate alternatives – defining a way to make a choice.

Problem formulation begins with a mental representation of the problem. However, due to the complexity and high degree of interrelatedness that may surround the problem, specialized problem formulation methods are frequently used to help structure and promote a systematic approach to the representation and manipulation of pertinent information. These methods specify how "images" of the problem are created, how information pertaining to the problem is examined and organized, and how that information is analyzed. Although formulation methods will differ based on how they emphasize the representation and manipulation of information, most fall into one of two general categories: formulation tools and formulation procedures (George, 1994). In either case, formulation methods tend to be more heuristic (based on judgment and "rules of thumb") than algorithmic and function to promote componential rather than marginal analysis. Examples of common formulation tools and procedures are listed in Table 1.1. In general, formulation tools provide a systematic way of representing information, whereas problem formulation procedures guide the process of manipulating that information to help clarify the problem. Ideally, these tools or procedures when carefully applied should improve the likelihood of making decisions that address complete and appropriate representations of the problem.

Today, geographic information systems have become an important formulation tool where facets of the problem can be visualized and sub-

Table 1.1 Selected problem formulation tools common to planning.

Tool	Description
Problem diagram	Arrows used to indicate the direction and nature of causal relationships between elements of the problem.
Decision graph	Decision areas are linked and grouped to indicate problem focus.
Interaction matrices	Nature and intensity of the interaction among factors, constraints, alternatives are displayed.
Q-methodology	Factors are uncovered through statistical analysis of problem elements.
Delphi technique	Grouped judgment is aggregated to form consensus.
Assumption analysis	Important and uncertain assumptions are identified, debated, and resolved.

jected to analytical operations that yield possible alternatives and solutions to a given problem.

Developing goals and objectives

If we can characterize planning and the process of designing a "vision" of the future, then to realize that vision we need goals and objectives to help focus our efforts and direct our actions. In general, planning goals reflect the ideological positions and social values of those involved in the process. Goals can be an affirmation of an ideal or a response to a problem, but in either case they are subjective and may change with time or circumstance. Within the planning process, goals provide direction first for plan-making and then later in the process for evaluation and decision-making (Kaiser et al., 1995). In general, goal-setting involves three interrelated activities:

1 Identifying present and future problems.
2 Determining community aspirations.
3 Identifying strategic issues and priorities.

In the context of these three activities, a goal represents an end toward which planning efforts are directed, while an objective is an intermediate condition achieved along the pathway toward some larger desired accomplishment (outcome). For example, a waterfront community may have expressed a goal to improve public access to the ocean shore. A policy to encourage the purchase of public right of ways may be an objective that will help realize the larger goal. Consequently, goals tend to be general in nature and expressed in broad terms. Therefore, we can consider goals to represent "broad brush" definitions of conditions that a community would like to realize though may never fully attain. Simple examples of goals expressed in a plan might include statements pertaining to:

- An enjoyable and safe environment.
- A well-balanced urban/environmental system.
- Preserving unique habitats.
- Providing maximum access to open space.

Five common types of goals expressed in plans have been summarized by Kaiser et al. (1995). These include:

1) **Legacy goals**: are left over from previously adopted and currently followed policies. Developing legacy goals begins with an inventory of the goals expressed in current plans. In some instances they may be inferred from patterns of past decisions or from an earlier stage in the current round of advanced planning. Using legacy goals as a starting point recognizes that a community has a history of discourse that defines its values.

2) **Mandated goals**: define requirements found in state or federal policy or from the judicial system's interpretation of statutory authority and constitutional rights. Such goals should be introduced into the community's goal-setting process to clarify directives that are important to the success of the plan.

3) **Generic goals**: describe ideals suggested by current thought and theory. Generic goals address matters of public interest on issues related to environmental quality, equity, quality of life, economic efficiency, and health and safety. Goals of this type may be viewed as an alternative source of community goals intended to support good practice and broader societal values.

4) **Community needs**: explain goals derived from forecasts of population, economic, and envi-

ronmental changes that require an appropriate response. Since planning is future oriented, forecasting is an integral part of the process. Translating forecasted changes in demand for housing, water supply, facilities, and waste disposal into needs allows future considerations to be balanced against other goals expressed within the community. Translating change into an expression of need involves the application of standards and a comparison of those standards against the projected future to be accommodated. Deviations from the standard help direct goals to meet desirable service requirements.

5) **Community aspirations**: characterize wants developed out of a participatory goal-setting process. The goals articulated by voices of the community define concerns and priorities that help focus and crystallize issues, problems, and desires as the community perceives them. Such goals help the planner gain a sense of what the public find important and what they see as their real needs.

Objectives tend to be much more specific when compared to goals, and prescribe steps that when followed produce attainable results. Expressed in concrete terms, the objective points to particular actions that can be implemented which, if followed, will produce a result related to the larger goal. These results can be measured or evaluated in relation to how successful they have been at bringing about an observable outcome. For example, a community may recognize that existing landfill capacity is low; therefore, to conserve capacity and extend the life expectancy of the existing landfill, a series of objectives may be proposed:

- Expand community recycling to include local business by a given date.
- Reduce the use of nonrecyclable products in local fast-food establishments by 30%.
- Develop capacity to accept a wider array of materials by 20% at local recycling centers.

Once goals and objectives have been established, they need to be examined and articulated in a form that allows the planning process to gain focus on their content. Making goals more than simple good intentions begins by

1 Ranking goals to prioritize their importance and provide a means to spot conflict.
2 Explicitly stating relationships between goals and objectives.
3 Selecting the most salient few objectives for a goal.
4 Examining the relationship between goals, their purpose, ends, and the means by which they can be pursued.

Each of these steps require information and a means to analyze data.

Data collection

With an understanding of the problem and a clear expression of the motivating goals and objectives, the next phase in the planning process requires the collection and synthesis of data specific to the goals. Data, information, and intelligence are essential for good planning. The question is what type of information and how much data is needed to produce it? Although there is no simple answer to this question, data collection and analysis actively direct our need to learn more about a given problem, its root causes, and to better understand the alternatives that may provide a solution. In essence, data supplies planning intelligence. It represents essential strategic decision support information that illuminates the problematic situation. The central problem in the data collection question, however, is that data is useless unless it can become information. Information in this regard describes a level of knowledge needed to solve a problem. Thus planning information should be able to answer in an accurate and timely fashion critical questions concerning:

- The nature of change.
- The pattern of opportunity and constraint.
- The important mitigating circumstances active in the planning area.

In this sense the data collected should be able to provide the information needed in order to facilitate the analysis of social, environmental, economic, and fiscal ramifications of change, and to compare trends to historic, current, and projected patterns (Kaiser et al., 1995).

In environmental planning there are "standard" items of data that have traditionally been

Table 1.2 Information and data applied in planning analysis.

Natural environment	Social/demographic factors
Slope	Population characteristics
Topography	Income patterns
Climate	Economic indicators
Vegetation	Employment patterns
Geology	
Natural hazard	**Governmental factors**
Hydrology	Jurisdictional boundaries
Wildlife habitat	Land development regulations
Air quality	Zoning regulations
Water quality	Tax rates
Noise	Annexation policies
Built environment	**Transportation characteristics**
Land use	Trip generation patterns
Road systems	Traffic volumes
Water supply systems	Road capacities
Housing stock	Modal characteristics
Viewsheds	
Historic structures/sites	**Indicator conditions**
Employment centers	Land ownership patterns
	Carrying capacities
Public facilities	Growth patterns
Schools	Decline patterns
Fire protection	Neighborhood characteristics
Police protection	Environmental quality trends
Libraries	
Churches	
Park and recreation	
Healthcare facilities	

considered essential to the process (Leung, 1989). A selection of data items commonly employed in environmental planning are summarized in Table 1.2. While they may not be indispensable, appropriateness and relevance to the problem determines if and when they become useful. In general, data availability, scope, and format impart the greatest influence on data collection efforts, and data gaps may be common for a variety of reasons.

The planning information base typically includes a mix of primary and secondary sources. Primary source material describes data collected from an original source. This may include surveys, air photos, satellite images, or data collected in the field. Secondary sources define data that has been collected and obtained by other parties and made available for use. Examples of secondary source data include census information and other documents, reports, or statistical tabulations assembled by local, state, or federal agencies. Depending on the nature of the problem, there are many types of data and collection procedures that can be employed, and care should be taken to ensure that the appropriate methods of data collection are used.

Identifying and selecting alternatives

Any goal or objective can be achieved in more than one way. Identifying and examining alternatives is an essential part of the planning process. Consideration of the alternative solutions to a given problem is important for several reasons. First, it suggests options that encourage debate and discussion regarding a given solution, its relative effectiveness, feasibility, and compatibility. Secondly, alternatives provide a basing point for raising questions about planning strategies, and the disposition of the motivating goals and objectives. Lastly, alternatives assist in the process of setting priorities in response to need. Unfortunately, the importance of generating alternatives has become a neglected dimension in recent planning theory (Bayne, 1995).

For any given problem a number of alternatives can generally be devised to meet a particular objective, and any one (or a combination of several) of these may be more appropriate than the original idea under the given set of circumstances. The issue confronting this phase of the planning process involves compiling a comprehensive listing of feasible alternatives. Developing that list places a premium on our understanding of the problem and the goals, and on our creativity. Creativity and thought are perhaps the two most critical influences when it comes to the task of conceptualizing alternatives. We have all heard the expression that "there is more than one way to skin a cat," but has anyone ever been told what those ways are? While much of the planning literature concentrates discussion on how alternatives are selected, evaluated, and compared, how planners come up with alternatives initially remains primarily an exercise in conceptual block-busting (Adams, 1974). Although a generic method cannot be offered, several techniques can be described that can create an atmosphere for thinking and exploration:

- *Brainstorming* – describes a group problem-solving technique that relies on creating an atmosphere of suspended judgment to encourage the articulation of ideas free of censoring. In a brainstorming session participants are asked to list ideas without concern for internal evaluation.
- *Synectics* – another group problem-solving technique fostering ideation, however unlike brainstorming, some evaluation is permitted. According to this technique, the problem is examined and restated to ensure that it is understood. Next, analogy and metaphor are used to allow participants to explore the problem in new ways. Finally, options are expressed that lead the group toward a solution.
- *Backcasting* – defines a method for exploring the implications of alternative development, the directions they move in, and the values that underly them (Robinson, 1988). Although not strictly a form of ideation, to undertake backcasting analysis, future goals and objectives are used to create a scenario of the future. This scenario is then evaluated in terms of its physical and socioeconomic feasibility. Iteration of the scenario is usually required to resolve inconsistencies and to mitigate adverse economic, social, and environmental impacts that are revealed during the analysis.

These examples suggest that alternatives emerge from creative thinking, an understanding of the problem, and a willingness to explore solutions that may challenge conventional "norms."

With a tangible set of alternatives listed, focus shifts from development to the question of selection. During the process of developing alternatives, little regard was given to the question of evaluation. However, when a set of alternatives must be analyzed and decided upon, greater emphasis is placed on concepts like feasibility, reasonableness, and the constraints surrounding each as judgment points help to narrow down the number of possibilities to a more appropriate list.

Table 1.3 Strategies and methods used to evaluate alternatives.

Method	Description
Matrix methods	Conducting a pair-wise comparison of alternatives against operational goals or anticipated benefits.
Linear programming	Quantitative evaluation of alternatives against a set of criteria variables to establish "best fit" relationship or optimal benefit.
Judgment trees	Evaluation of causal interaction between alternatives and anticipated or desired outcomes based on using judgment or subjective probabilities to derive best solution.
Scenario analysis	Placing alternatives into a description of a desired future state and evaluating either qualitatively or quantitatively how selection of a given alternative may influence the future.
Simulation	Developing and using a model to explore the relative impact or success of an alternative and evaluating the "what if" ramifications of a given selection.

The analysis of alternatives defines the general process of determining the effects or impacts of each against a goal or objective in question. This phase of the planning process cannot proceed, however, without the choice of criteria for making evaluations. Alternatives are typically evaluated both qualitatively and quantitatively and assessed in relation to their physical, social, economic, fiscal, environmental, and aesthetic implications on the planning area. A wide assortment of tools and techniques has been devised to assist with alternative analysis and selection. A sample of common approaches to the issue of selecting alternatives is presented in Table 1.3.

The methods identified in Table 1.3 focus on two major evaluation tasks: forecasting and comparison. Forecasting the impacts associated with a given alternative can be accomplished in one of three ways: (1) *extrapolation* – the extension of a trend logically into the future based on past behaviors; (2) *modeling* – the creation of a representation of the situation that can be examined; (3) *intuition* – the application of judgment and experience. The principal features of these approaches have been reviewed by Sawicki (1988).

- **Extrapolation** – trend extrapolation is based on the empirical examination of some phenomena with respect to measurements taken across time. Forecasting through the projection of trends is a frequently used method of exploring future conditions. As a technique it may include the use of moving averages, linear regression, curvilinear regression, or envelope curves to fit a line to the data points that summarize the important trend. An excellent discussion of trend extrapolation can be found in Hill (1978).

- **Modeling** – a model is simply a representation of an object, system, or concept in a form different from the entity itself. All decisions are made on the basis of some type of model, whether a formal computer representation written in a programming language or a simple "idea" of how we think something works or behaves. Models provide a means to simplify complex problems. Approaches to use of models in a planning context have been discussed by Gordon (1985) and Lein (1997).

- **Intuitive forecasting** – implies the use of expert judgment and experience to forecast the possible outcome of an alternative action. The use of judgment and informal heuristic reasoning is a type of knowledge acquisition process where the analyst queries a group of experts to illicit a causal process related to an alternative. Perhaps the most widely practiced form of intuitive forecasting is the Delphi method. This approach is described in detail by Linstone and Turoff (1975). Lein (1993b) has examined other forms of formalizing judgment for application in environmental forecasting as well.

Comparison methods describe a family of techniques designed to facilitate the "ranking" of alternatives or to provide relative measures of attractiveness that can be used to prioritize and select among alternatives. Commonly used methods of comparison include matrix methods, scaling techniques, and programming designs.

- **Matrix methods** – a matrix describes a two-dimensional system of rows and columns that allows pair-wise comparisons of alternatives against evaluative criteria. Because of its two-dimensional structure, a matrix provides a tabular format that simplifies one's ability to visualize the interactions between alternatives. Within the cells of the matrix, symbols or scores can be assigned to identify critical relationships and possible conflicts that can be associated with a given alternative. The matrix can then be used to first identify effects by systematically checking each alternative against the criteria set, and secondly to ascertain the relative importance or significance of the effect. Should an impact become evident, a "score" is placed in the corresponding cell. Although scoring implies numerical measurement, in actuality scores suggest subjectively derived evaluations that are employed to express or rate the relative attractiveness of an alternative in pseudo quantitative terms.

- **Scaling techniques** – scaling or rating methods are based on the assumption that an attractive score (*S*) can be derived for a set of (*i*) alternatives using the general relationship

$$S_i = \Sigma \kappa_i v_{ij}$$

where S_i equals the total value of scores for alternative (*i*), κ_i explains the weight placed on criteria (*j*) and v_{ij} defines the relative value achieved by criteria (*j*) for alternative (*i*). This fundamental relationship has been extended to create a range of multicriteria, multi-objective decision aides that are useful in situations where more than one criterion is needed to asses the attractiveness of an alternative solution. Perhaps the best-known and most widely adapted of these was introduced by Saaty (1977). The main objective of multicriteria scaling techniques are: (1) to identify choice alternatives satisfying the objectives in relation to the problem, and (2) to reduce and order the set of feasible choices to the most preferred alternative.

- **Programming designs** – programming designs apply mathematical or statistical procedures to select the optimal allocation of resources needed in order to achieve the desired goal with a minimum of "cost." Programming designs require identifying a set of decision variables, criteria for choosing the "best" (optimal) values of the decision variables, and a set of constraints or operating rules that govern the procedure. These terms are expressed in the form of linear equations or linear inequalities written as functional relationships of the decision variables.

Regardless of approach taken to formulate and select alternatives, there are four basic principles that guide the process (Sawicki, 1988):

1 Conclusions drawn about each alternative should be displayed in a way that is simple and transparent.
2 Techniques used for selecting alternatives must be capable of handling multiple criteria, and the advantages/disadvantages and trade-offs made visible.
3 Consideration must be given to nonquantifiable criteria and methods to include these criteria should be utilized where possible to accommodate their evaluation.
4 Methodologies should lend themselves to a decision and display the attributes of each alternative to allow consideration of all relevant factors and permit compromise when that becomes necessary to reach a decision.

Synthesis and implementation

The final phase of the planning process described in this chapter is an amalgam of elements needed to make planning work. A major concern in planning is that all too often the process becomes the central focus and the "plan" becomes a document that resides on a shelf with little hope of becoming realized. With goals and objectives carefully articulated, data collected and analyzed, the problem well defined, and a set of alternatives selected, the plan begins to take shape, not just as a document, but as a well-integrated idea. Integration implies synthesis with an eye toward the future. In this context, synthesis is concerned with the degree to which elements of the problem and solutions fit into a framework for action. It takes creative thinking and critical evaluation to create this framework and to ensure that the plan will encourage good decisions. The questions of creative thinking and critical evaluation draw attention to the plan itself and how this plan relates to the future.

A plan may be conceptualized in a number of ways. In one sense we may think of a plan as a "blueprint" for the future. As such, the plan becomes a detailed documentation of the environmental characteristics, community features, problems, goals, objectives, recommendations, and programs germane to design of a desired future state of the community. Viewed in this manner, the plan becomes a statement of policies that explain what the community wants to achieve relative to its environment (physical, social, economic, aesthetic) and a physical document with specific language to illustrate, educate, and direct the design of this future.

Although the technical content of a plan can vary, certain elements are commonly included:
- Introduction and background to the plan.
- Statement of purpose.
- Description and documentation of the planning area.
- Elements of the plan.
- Statement of findings.
- Recommendations and evaluation.
- Implementation strategies.

Implementation has been described as one of the more difficult phases of planning. A major reason for this is that implementation moves us from the "science" of planning to the political realities in which planning operates. Since implementation identifies the actual carrying-out of the plan and its recommendations, implementation must enable the outcome. This enabling aspect of planning may require bringing together the necessary legal instruments, policy mandates, or building existing law and programs into the plan as part of its implementation. This suggests that plan implementation may proceed in either of two ways. One way calls simply for the adoption of the plan, letting its policy recommendations become translated into design and policy actions. Taking this approach may create conflict and uncertainty. Consequently, the plan may be adopted as a pilot program or demonstration project prior to full implementation. By so doing, the implementing agency has the time needed to acquire experience with the program, monitor its effectiveness, and adjust the program where necessary prior to widescale adoption.

A similar strategy may call for a phased implementation where certain elements or recommendations of the plan are adopted according to a timing schedule. Phased implementation may be appropriate when the plan requires specific legislative action, or to give those affected by some aspect of the plan critical time to prepare. In either case, implementation requires a program that adequately addresses the issues which may hinder realization of the plan. With a sound implementation program the number of obstacles encountered can be kept to a minimum. However, new problems will arise that will require repeating or revising earlier phases of the planning process. Recently, Talen (1996) has reviewed the implementation problem and offered a typology for plan evaluation, yet the planning process and all the elements that it embodies are complex. Keeping all the features of the plan and all of the factors that need to be included requires a management strategy. In the following section systems analysis is examined as one possible strategy the environmental planner can call upon to help organize the intricacies of the problem.

Adopting a systems view of planning

Planning is a complex task simply because the subject matter involved is multitemporal, multivariate, and multidimensional. To make the planning process work requires not just an organizational framework of tasks or phases, but a construct that integrates elements of the problem into a synoptic view (the big picture) that facilitates understanding, guides analysis, and supports prediction. One useful construct for organizing the complexity of the problem in a manner that enhances understanding of interrelatedness and interdependence introduces the concept of a system and the methods of systems analysis (Chadwick, 1974; Feldt, 1988).

The concept of a system has been used in a wide range of contexts. In some instances, its meaning may be implicit in how it is being used. However, because it is a concept that is so widely applied, a formal definition will help connect us to the planning problem. A system may be defined in several ways (Lein, 1997). Perhaps the most fundamental explanation of the concept characterizes a system as a set of objects together with relationships between the objects and their attributes (Hall & Fagen, 1959). Put another way, a system is nothing more that a set of interrelated elements together with relations between the elements and among their states that function in a complementary manner. What is important about these definitions is that we can extend the concept beyond the idea of a physical entity and describe a system as a perspective and a subject of inquiry. Considered in this way, a system becomes a "model" that represents a way of thinking about how things are connected and how they work, whether we are talking about space stations or planning areas. A system also becomes a way to organize the complexity of observed reality and somehow manage or control that complexity. In addition, the system concept encourages a functional view of the real world and helps us recognize that there are purposeful connections that bind elements of a problem together into a coherent structure. One need not be an automotive engineer to understand how an automobile's cooling system works. When a car overheats and stops running, we can understand the problem by a basic model of the car's cooling system. Similarly, we can appreciate the relationships that make up the planning area by casting them into this same "systems" framework. So watershed and neighborhood can become systems and we can examine what they are made up of, how they work, and, more importantly, how they change.

From the planner's perspective, perhaps the most important quality of systems thinking is that it directs our attention to the "whole" and fosters identification of cause and effect processes. This concern for process may lead to prediction and the representation of events that make a possible future discernible. Although prediction may not be the goal, the system model lends itself to analysis and simulation. Therefore, the value of defining and analyzing systems is that they enable the structure and behavior of complex interrelationships to be explored (Bennet & Chorely, 1978). Consider the example of land use. Using the system concept, the relationships between types of uses, their location, arrangements and juxtapositions can be placed into a model that allow for some understanding of the patterns visible on the landscape. With this model, the complexities of how residential uses support commercial areas, and the mix of uses needed to maintain economic efficiency and serve a given level of population can be examined. In a similar vein, the system concept can be used to explain the interaction between urban processes and the environment, and depends only on how that interaction is defined. Definition in this regard is the key. Because a planning problem may be complex, there is a natural reaction to isolate parts of the problem and explain how each of these parts operate under simplified conditions. For this simplification to work, these isolated pieces of reality must maintain connection with the real-world. To maintain this connection a method is needed. This is the method of systems analysis (Huggett, 1980).

As an analytic device, a system can be defined at varying levels of resolution and detail (Lein, 1997). One type of system we can identify is the abstract system. With an abstract system, the elements comprising its structure are concepts whose components have connecting relationships based on certain assumptions. A second type of system is referred to as a concrete system, where at least two defining elements are actual objects. When we apply systems methods in planning, we are also interested in how to represent change and capture the dynamic nature of the problem using the system design. To conceptualize and capture process, the list of system types can be expanded to include other relevant forms. These include:

Static systems – systems whose states are held in equilibrium conditions.

Dynamic systems – systems whose states vary over time.

Homeostatic systems – systems that strive to maintain balance.

Continuous systems – systems that display behaviors uninterrupted over time.

Discrete systems – systems where change occurs in finite time intervals.

Stochastic systems – systems whose behaviors are influenced by an element of randomness or chance.

Deterministic systems – systems where future states are dependent upon direct functional links to past states.

These representations provide a focus in the system design process that leads to the formulation of a model. As Lein (1997) notes, through the application of these system concepts and the formulation of a process-oriented system design, the complexity which surrounds a planning problem can be reduced to an ordered and structured set of objects that helps us "see." Producing this model depends on the success to which the system has been defined. To the planner this involves four critical steps:

1 Specifying the variables to use in the system.
2 Stating the hypothetical relationships that define variables comprising the system.
3 Developing a simple explanation of the system and its structure.
4 Testing and refining the system model.

Adopting a systems view of planning compliments the planning process outlined previously in several ways. Both require recognition and definition of the problem, a set of goals and objectives; and once the model has been developed it must also be implemented. A general outline explaining the steps followed when performing a systems analysis is given in Fig. 1.4. As suggested by the illustration, systems analysis begins with the critical step of identifying the components that will define the system.

Defining the system and possible subsystem components is based upon several presumptions regarding relationships that will organize and connect elements together. It is also typically assumed that a meaningful structure can be hypothesized that separates the system from the real world. Taking these ideas and applying them to planning we will note that systems will possess

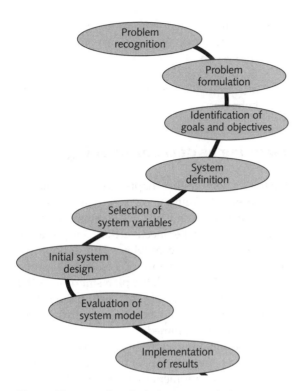

Fig. 1.4 The general method of systems analysis.

two important features that must be identified and described in order to produce something meaningful: (1) A functional or process structure that characterizes some definition of a flow and (2) a morphological structure that defines a spatial (geographic) arrangement. These two features help direct our inquiry and assist us in adding specifics to the model that make its representation more useful. Specifics draws attention to the details of our design and the requirements that the features embedded in the system provide a dynamic view of the problem. Several key attributes assist us in developing this type of representation. Among the more relevant to the application of systems methods in planning are:

- **System state** – the state of a system at a given point in time is the set of properties it described (i.e. number, size, age, color, mass). The system state is defined by the value these variables have at that instant in time.
- **System environment** – because every sys-

tem is composed of elements, systems are bounded into an environment space. This environment is a set of elements and their relevant properties that are not part of the system but can influence its state. Therefore, a system environment consists of all variables that can affect its state. State variables outside the system as bounded are termed exogenous variables, while those within the system are called endogenous.
- **System interaction** – a system that is defined in such a way that no interaction takes place with elements not contained within it (completely self-contained) explains a closed system. Conversely, a system which displays interaction with its environment is an open system.
- **System event** – characterizes change in one or more properties of the system. This change will occur over a period of time with a specific duration.

Systems, change, and feedback

Among the more useful aspect of systems analysis is that it offers a perspective from which the planner can study change. Because planning has been described as a future-oriented activity, projecting, predicting, and responding to the changing status of the planning area is of fundamental interest to the planner. In planning we are constantly asked to explain change, describe the processes that drive it, identify its consequences, and predict the behavior of systems subjected to change. When examining change our attention is directed toward dynamic systems and their characteristics. Since change may be said to manifest as a deviation in system state, it can be observed by noting the disposition of the system's state variables. Through careful observation cause and effect relationships can be categorized in one of three principal ways:

1 Reaction – a system event that is deterministically caused by another event.
2 Response – a system event produced by another system or environmental stimulus.
3 Behavior – a system change that initiates other events.

Another important concept when system methods are applied in planning is that of feedback. Regardless of type, any system operating in the environment exhibits a degree of sensitivity to the manner by which its defining components are connected and arranged. Loosely defined, feedback explains the return of information as input to the system. This cycle of returned information acts directly on the performance of the system and its structure. Feedback can assume two basic forms: (1) positive feedback – characterizing a "deviation-amplifying" process that influences a change in state and functions to maintain that change, and (2) negative feedback – characterizing a "deviation-dampening" process that retards the effects of change in the system.

Specifying the design

With the fundamental features of systems analysis understood, a basic design of the problem as a system can be produced. For the systems approach to work, however, we must have a hypothesis or process around which design can focus. This nucleus of our system can be a very general idea or a specific relationship where representation in system form will enhance understanding. To achieve this important goal the design qualities need to be arrived at so that a function model of the problem may be realized. The critical factors influencing the design on a system are:

- Size – the number of variables that comprise the system must be determined with special interest given to those controlling variables that exert the greatest influence over its behavior.
- Associations – this property specifies how the variables relate to one another, where consideration is given to the degree of correlation among variables, and the strength, direction, and sensitivity of those relationships.
- Causality – the connection of cause and effect to process, how process directs the system, and the manner by which it functions.
- Pattern – the intercorrelations defining the network that connects elements together in the system.

- External forcings – the influence of an external variable on one or more components of the system.
- Inputs and outputs – the nature of flow into and from the system that gives rise to its behavior.

Planning as decision-making

Up to this point in this chapter we have explored the concept of planning, the basic features of the planning process, and the role of systems methods as an organizing perspective that can guide us through the planning problem. Throughout, we have suggested that planning is essentially a type of decision-making with the plan standing as the primary decision focus. In this section the nature of decision-making and its connection to the planning problem is examined.

Decision-making has been defined as a process by which a person, group, or organization identifies a choice or judgment to be made, gathers and evaluates information about alternatives, and selects from among those alternatives. These familiar steps paint the process of decision-making as a stream of thoughts and behaviors that also include elements of risk and uncertainty (Lein, 1997). The decision, in this context, unfolds through the combination of learning, understanding, information processing, and information accessing, all with careful definition of the problem and the circumstances involved. This simple description of the decision-making process is underscored by a reasoning method used by people when approaching a decision problem. Because everyone approaches a decision differently, it is critical to understand how such factors such as personal background, experience, inherent psychological conditioning, and the situation surrounding the problem will influence both the way decisions are made and how the problem is perceived. To understand how these factors direct the process of deciding, several models have been offered that introduce and summarize the various styles decision-making can take (Davis, 1988).

- **The rational model** – we have discussed this approach with respect to the planning

THE NATURE OF PLANNING **19**

model. This decision style views decision-making as a structured process where a group systematically reduces the decision problem to a set of measurable quantities or qualities. The comparative merits of each determine a possible outcome, and that alternative with the greatest merit (value) is selected.

- **The organizational model** – decision-making according to this model follows the established policies or guidelines of an organization. Here, the decision-maker takes action based on a set of guidelines or policies rather than evaluating the relevant factors that influence the decision. In this regard, the decision-maker avoids uncertainty by following a predetermined path.
- **The political model** – places decision-making in a political setting where decisions result from group interaction and deliberation. In this setting individuals rely on persuasion or authority to satisfy subjective goals. According to this model, there is no universally accepted best decision, but rather the identification of an alternative that provides the most acceptable solution.
- **The satisficing model** – recognizes that an optimal solution to a problem may not exist. In such instances, the decision-maker seeks an adequate alternative, one that satisfies one or more initial requirements of the solution. From here, the decision-maker relies on feedback to improve the next iteration of the problem.

Realizing that a problem will be approached differently depending on circumstance, and that many factors direct planning decisions, gives emphasis to the setting in which decisions are made and how setting may influence or shape a solution. Therefore, while certain problems may be routine and repetitive and conform easily to an organizational style of decision-making, it is more likely that planning problems will be unstructured, unique, and require the exercise of judgment, intelligence, and adaptive problem-solving behavior. Such nonprogrammed decisions may initially be approached using the rational model. However, it is more likely that constraints im-

posed by time, situational factors, and financial limitations will move decision-making toward a more satisfycing mode. These constraints point to the many issues that propel planning and influence how the planning process unfolds. Several of the more relevant issues are examined in the next section.

Propelling issues in planning

Planning focuses on the management and maintenance of the human landscape: that mix of social, cultural, economic, political, and administrative attributes that reflect who we are and what we deem important and essential to our survival and sustenance. The issues of planning are as basic as the needs of human beings multiplied many-fold. These fundamental planning issues are no different today at the beginning of the twenty-first century than they were 100 years earlier. We can consider these as the constants of planning which explain the fundamental needs of a society:

- the provision of housing,
- the allocation of employment sources,
- the facilitation of commercial and service functions,
- the delivery of clean water,
- the removal of solid and liquid waste,
- the production of a healthful environment,
- the design of efficient means of communication and transportation,
- the creation of a functioning and balanced system of institutions, and
- the maintenance of recreation and aesthetic qualities within the built environment.

Yet each item listed above suggests a larger context composed of numerous driving forces that propel change, moderating forces that redirect and transform change, mitigating forces that provide balance and a resource base from which the material to sustain change are derived. Through the interplay of these complex actors the human world – the urban environment – takes form. The nature of that form, its magnitude, and consequence both in human terms as well as those of a geographic entity that interacts within the boundaries of a greater environmental system move the planning problem to a higher plane of awareness and purpose. It is here that the distinctions compartmentalizing planning into specific specializations begin to blur and the planning problem broadens in perspective.

At this level the issues of planning may appear to be more conceptual, yet they remain intertwined with those fundamental needs of society and connected to the patterns those needs display. Here, urban form becomes the driving force of change that exerts its own influence in both social and environmental terms. The recognized and potential impact of these changing patterns, coupled with the demands it places on its primary means of support and the manner by which these demands are satisfied, introduce an entirely new set of concerns. Four propelling concerns that form the backdrop against which all planning issues will be framed follow.

1 Urban growth and growth management

The development of urban form is an attractive force. As human habitat it offers the support systems that enhance our quality of life. As we become attracted to the possibilities offered within this landscape, urban form multiplies. This growth is due in part to the development pressures encouraged by our own demands. Thus as our demands become realized and satisfied, built form expands. Such expansion increases wealth, which in turn encourages new demands, which become satisfied through yet another round of expansion. Expansion is both a physical quality that assumes a geographic expression and a social feature explained in terms of increases in population, exchanges, variety, opportunity, and preferences. At the opposite end of the growth question are the management issues that must keep pace with each cycle of expansion. These issues are not well articulated in the market forces or the opportunity/ demand preferences that fuel growth, but they are there. Consider the simple question: What does it take to keep the urban system functioning? Addressing this questions directs our attention to the energy, water, waste capacities, infrastructure, and other support services that those of us living in this landscape rely on. With each round of expansion those support functions must also expand. It has been noted that the urban system grows cheaply, but is expensive to maintain. The diminishing returns of growth, the vexing problems associated with solid and hazardous waste removal, land-use change, congestion, sprawl are all features of the management problem. These features also place in sharp contrast the competing realities of the built environment as a consumption system, commercial system, production system, and environmental system (Douglas, 1983). Reconciling these differing views and roles is one aspect of the growth management problem. Put simply, how does one maintain human habitat when human habitat is dependent on natural habitat for its survival, and what are the costs involved?

2 Sustainable development and cumulative change

Growth is not an evil, although it is easy to couch it in those terms. Growth is a reality that we depend on to maintain our livelihood. The question, therefore, is not the simplistic dichotomy of growth or no growth, but rather the redefinition of growth in more sustainable terms. The concept of sustainability has become a much overused term in the debate surrounding the question of growth and environmental change. While a basic definition of the term is insufficient, sustainability is a concept with many implications. First the concept suggests a wider view of the growth and development process. Traditionally, growth is described in economic terms and tends to assume a human-centered perspective. From this point of view the physical processes that feed growth and supply resources to support the built environment tend to be poorly integrated into the development model. A sustainable view is more integrative and places human landscape as part of the fabric of a larger environmental system. A second aspect of this concept directs us to adopt a much longer time horizon toward what we consider the future than is typical in most planning applications. Thus, rather than the usual 1- to 5-year planning horizon, sustainability calls for a conceptualization of tomorrow that spans several generations into the future. Another implication of the term influences how we view and use the resources needed to maintain our built environment. Sustainability is based on the use of renewable and perpetual resources in harmony with the ecological system that patterns the landscape. Such a view stands as an important compromise between the extremes of no-growth versus unlimited growth, and directs planning to consider alternatives that promote efficiency and environmental balance. Perhaps the most important implications of sustainability are societal. Sustainability is a model of social, economic, and environmental interaction that will foster change in the manner by which basic human needs are met. With respect to planning, a move toward a sustainable system will require critical transitions in our political, economic, and resource systems, and essential ethical and behavioral shifts in attitude toward a reshaped worldview that removes the separation of humans and environment.

3 Equity distribution and conflict

As the forces of growth and change commit society to a more intensive use of the earth's surface, the question of fairness and the avoidance of conflict will exert great influence on the planning process. With heightened awareness and concern over the impacts of growth and the implications of environmental change on all corners of society and the ecosystem, the simple solutions and rationales of the past are likely to be ineffective and unacceptable to a growing segment of the population. Environmental equity has recently emerged as a critical issue. It developed slowly from the observation that socioeconomically disadvantaged groups historically bear a disproportionate burden of risk and hazard in the environmental policy-making arena. The equitable treatment of all races and cultures in decision-making represents a challenge to the practice of planning and the processes by which planning decisions have been made. Through greater efforts to improve citizen participation and the inclusion of advocacy groups, mediation and conflict resolution strategies may redirect how plans are made.

4 Environmental process and urban entropy

Focus on the human landscape and the economic and social processes that motivate and shape it has traditionally ignored the environmental processes modified as a consequence. Bringing the environment more directly into the planning process is an idea nearly three decades old, yet the renewed emphasis on maintaining balance within the built environment point to the importance of the natural processes that regulate and ultimately control the scale and extent of human endeavors. Integrating environmental processes into the planning process means more than simply understanding the physical characteristics of the planning area. It means understanding how the environment works and recognizing its potentials, limitations, and risks as active elements of our planning efforts. The total environment and the interactions that describe the form and function of the landscape create the need to develop broader goals. These goals begin with the requirement to plan in close accordance with natural processes and culminates in a reinterpretation of the built environment not as an artificial arrangement superimposed on a natural system, but as a synthesis of processes that create a recombinant form. In many respects this issue is intimately linked to the ideas presented throughout this chapter. In addition, they provide the theoretical focus that bridges the entropic effects of the growth model with a sustainable perspective based on a wider inclusion of stakeholder interests and a tighter integration of natural process as an organizing structure from which a vision of the future can be assembled.

Summary

Planning is a fundamental human activity with common themes and problems. In this chapter the concept of planning was examined and dissected. Discussion introduced the logical process and methods common to planning to demonstrate the formalisms associated with this type of decision-making. Whether explained in ordinary terms or with reference to a technical specialization such as environmental planning, the process begins when goals and objectives are established, information is analyzed, and alternatives are compared, and culminates when a course of action is selected. Knowing what is needed to organize thinking, and understanding whom the plan is for and when optimal solutions and appropriate alternatives are found are complex questions that can be difficult to answer. In this chapter a framework was presented that relies upon the methods of systems analysis. Systems analysis is a well-tested tool for managing complexity. Through its application, the planner may approach complex problems, organize thinking, and form an understanding of problems that will enable better solutions to emerge.

Focusing questions

Explain the concept of planning.

Discuss the rationale that supports the planning process.

What are goals and objectives, how are they derived, and how to they help define the "right" planning problem?

Critically evaluate the utility of systems thinking in planning and explain how process-oriented thinking drives proactive decision-making.

CHAPTER 2

Defining the Environmental Approach

The environment conveys many different meanings in planning. At one level the natural environment supplies the land required to accommodate growth and development. At another, the environment describes a set of resources to draw from and conserve. At still another level, the environment defines a series of natural functions to be maintained, hazards to be avoided, and opportunities to be exploited. Finally, there is the view of the natural environment as an all-encompassing entity that simply exits not as a passive feature there to serve human needs, but as a set of active processes that define a behavior and establish patterns that interact with and redirect human trajectories. The complexities, conflicts, and contradictions inherent to these contrasting explanations of the term "environment" present a challenge to planning and force the planner to look beyond the immediate dictates of land markets and the goals motivating economic growth. The environment demands wider consideration of all the relevant factors that drive planning and shape the landscape.

It was the recognition nearly four decades ago that wider environmental considerations needed to be incorporated into plan-making that contributed to the evolution of environmental planning. Today that recognition remains a top priority, and continues to broaden the scope and purpose of environmental planning and the role the environment plays as a decision criterion in the planning and development process (Honachefsky, 2000; Leitman, 1999). To appreciate the significance of environmental planning as both a school of thought and a professional practice, we need to explore precisely what it means, what differentiates this approach from the more traditional forms of land use and urban planning, and how one performs planning according to this theory.

The nature of environmental planning

If we take all the various definitions of planning offered in the previous chapter and assemble them into one simple statement, we can characterize the purpose of planning as the process of allocating functions to their appropriate spatial location. So stated, this basic explanation poses a question that provides an important point of departure from traditional planning approaches and suggests room for an alternate strategy. Looking carefully at the definition offered above, the notion of allocating functions to an appropriate spatial location asks us to consider what is meant by the word appropriate. In the majority of instances, our response to the question would center around a set of common decision points: economic rationality, efficiency in the provision of services, accessibility with respect to population, attractiveness in relation to amenity or aesthetic qualities, feasibility considerations with respect to engineering and construction concerns, and acceptability given the local political landscape. Using these decision points, a site becomes appropriate for a given development proposal when it falls comfortably within established parameters for categorizing each criteria as such. Of course, there are alternative definitions of the term "appropriate," and one critical definition absent from the above is an environmental definition. Consideration of appropriateness from an environmental perspective opens a gap for environmental planning to fill.

Utilizing an environmental rationale, appropriateness directs us to consider, in addition to those criteria listed above, the **compatibility** of the proposed function within the fabric of the ecological system, its **suitability** in relation to the physical and environmental qualities of the site, its **susceptibility** with regard to the potential environmental impacts, and its **sustainability** as explained relative to the long-term functioning of environmental processes and the maintenance of environmental integrity. These considerations and alternative views of the concept create the need to maintain balance between the productive use of the land and natural resources and the maintenance of ecological functioning. Approaching this balance and recasting "appropriateness" considerations with greater emphasis on environmental criteria is the fundamental goal of environmental planning.

While a succinct definition may be elusive, environmental planning as an activity involves the use of biophysical and sociocultural information to suggest opportunities and constraints in relation to land development in a manner that seeks to explain the fitness of the environment to support a given function. A more comprehensive definition of the concept has been offered by Baldwin (1985). According to Baldwin, environmental planning may be defined as the initiation and operation of activities to direct and control the acquisition, transformation, distribution, and disposal of resources in a manner capable of sustaining human activities with a minimum disruption of physical, ecological, and social processes. Although definitions will vary, the environmental approach to planning seeks to explore economic growth alternatives that are socially and environmentally sustainable.

The overriding goal of environmental planning involves balancing human needs *with* the dynamic properties that constitute the environment. In fact the concept of balance is so central that is rests at the core of everything the environmental planner does (Holling, 1978; Westman, 1985; Margerun, 1997). However, balance is a difficult quality to achieve for a variety of reasons, and thus remains a challenge that propels the evolution of this form of planning (Baldwin, 1985). The factors that frustrate the balancing of human needs with environmental quality include:

- The complexity and interrelatedness of environmental problems and solutions.
- The evolving nature of environmental planning knowledge.
- The frequent omission or discounting of environmental goods and services during conventional value analysis.
- The difficulty of achieving environmental goals that require significant changes in lifestyle.
- The conflict between environmental goals and community development goals.
- The difficulty in establishing environmental priorities and defining trade-offs.
- The lack of consistent commitment of resources to environmental programs.
- The general lack of information and support tools needed for sound environmental decision-making.

While achieving such balance is a challenge, environmental planning theory and practice has produced a variety of approaches to the formulation and implementation of solutions to meet that challenge. Each approach reflects a particular philosophy or mode of analysis regarding the environment and how environmental problems may be conceptualized, defined, studied, and solved (Briassoulis, 1989). To begin this discussion of environmental planning strategies, it is instructive to first review the philosophical foundations that influence environmental planning thought.

Philosophical antecedents

Environmental Planning describes a specific set of views strongly influenced by a lineage of ideas, beliefs, and values that color our present-day perceptions and practices. The evolution of modern environmental thinking has been examined extensively by Pepper (1984) and Ortolano (1984). From these reviews we can conclude that our web of environmental beliefs takes shape based on a set of some very fundamental principles (Buchholz, 1993):

1 **Conservation as the efficient use of resources** – this principle recognizes that using the natural environment to satisfy material needs is a necessity. However, resources are scarce and consideration must be given to their efficient use and the minimization of waste.

2 **Maintenance of harmony between people and nature** – the impact of human actions on the environment forms the basis for this ideal. As human activities introduce disruptions that effect natural systems, there is a need to maintain balance between people and nature and preserve the integrity of natural systems.

3 **The environment as spiritual renewal** – the beliefs expressed under this heading define the environment in moral, religious, and aesthetic terms. Taken together, these ideas characterize the environment as an entity possessing ethereal qualities that are worthy of preservation and protection simply because they exist. While these beliefs may be seen purely as philosophical argument, they are deeply ingrained in cultural values and suggest that the environment is seen as more than a source of resources, but also a source of renewal, recreation, and amenity.

4 **Natural rights and the shared existence of humans and nature** – the ideals expressed here identify the belief that natural objects and nonhuman animals enjoy "rights" that ensure their existence and survival. Central to this set of environmental beliefs is the idea

that humans "share" the environment and are bound by an ethic that should influence the way we make choices that will affect whether or not the biotic community will remain a viable habitat apart from human occupation.

These ideals form the basis of our environmental consciousness and serve as the foundation for understanding contemporary environmental attitudes. Although when one is confronted with an environmental problem a combination of ideas and attitudes will be expressed, the voices of concern typically align themselves with positions

• that hold to principles of conservation and the wise use of resources;

• that maintain the need to control human actions and reduce irreversible impacts on natural systems;

• that ascribe aesthetic and spiritual qualities to the environment;

• that appeal to ethical principles calling for restraint on human actions that adversely affect the environment.

Perhaps the most significant aspect of our environmental web of belief is that it is subject to change and made complicated by variations in personal backgrounds, perceptions, and values. Presently, the dominant environmental paradigm views all environmental issues in a global context rather than expressing them within a regional or local framework. At this global scale, environmental questions are highly interrelated, foster collective action, and involve much broader agendas than simply those of economic expansion. Thus while the contrasting expressions that define environmental thinking may appear confusing, the underlying concerns on which they are based have been translated into ethical norms and governmental policies that guide the way decisions affecting the environment are made (Ortolano, 1984).

From an exclusively human-centered orientation, decisions affecting the natural environment couch the problem of selecting among alternatives in terms of one of several evaluative strategies. Several of the more relevant to environmental planning have been examined by Ortolano (1984) and include:

- **Utilitarianism** – a decision-making framework based on estimates of the beneficial and harmful consequences of a policy to society as a whole. The principle of utilitarianism is based on the premise that decision-makers should select that alternative that produces the greatest net balance of beneficial over harmful consequences to society. Adopting a utilitarian perspective wil encourage judging an action entirely by its outcome to society in total. This approach carries the implicit assumption that harmful and beneficial consequences of an action can be predicted and evaluated in terms that can be equated with a net effect that provides evidence that satisfactorily weighs benefits against costs.

- **Cost/benefit rationale** – a more formal adaptation of the utilitarian strategy. The significant departure of cost/benefit decision-making is based on defining beneficial and harmful consequences in monetary terms. Accordingly, social benefits of an action are expressed relative to the amount of money individuals would be willing to pay to obtain the beneficial consequences, while the social costs are measured in relation to the opportunities society gives up when its resources are used to implement the proposed action. To the decision-maker the criterion for choosing among alternatives simplifies to the selection of that alternative with the greatest numerical difference between monetary benefits to costs. The principal limitation with this approach and a continuing source of controversy frustrating its application centers around the problem of (1) equating human actions in monetary terms, and (2) relating those actions to intangible environmental qualities. This has contributed to the use of qualitative comparisons that enable a more systematic evaluation of alternatives (Swartzman, Linoff, & Croke, 1982).

- **Equity distribution** – applies principles of fairness to decisions affecting environmental quality. Equity distribution is concerned with the just allocation of benefits to costs and attempts to reconcile the disparity that while some individuals enjoy environmental benefits others must incur a disproportionate level of costs. Although the concept of fairness has received increased attention, the analysis of equity issues is frequently confounded by problems in measurement and by concept inconsistencies.

- **Rights to a habitable environment** – as a basis for decision-making, this approach directs the decision-maker to consider whether an alternative will affect the moral rights of humans and other living things to an environment that they can exist within. In this context, a moral right is one that is independent of any legal system and is based more on moral norms than legal definitions. While the notion that humans have a right to a livable environment may seem self-evident, extending this principle to include a wider interpretation of habitat directs the decision-maker to consider alternatives based on their ability to maximize the viability of habitat for the range of species that may be affected.

- **The context of future generations** – no other concept cements environmental thinking as much as the question of future generations. Because planning decisions frequently set into motion a series of irreversible and irretrievable commitments of land and other resources, the context of future generations directs the planning problem beyond the immediate concerns of the present and forces consideration of the implications of a decision that may (1) foreclose on future options or (2) introduce risks that future generations will have to confront. While the question of whether future generations have a moral right to an environment that has not been depleted of its resources remains in debate, the idea that each generation is a trustee of the environment for succeeding generations has placed an important role in shaping environmental policies (Smith, 1995).

The strategies identified above, together with the primary avenues of thought pertaining to the environment, remind us that the environmental planning problem can never be addressed on the basis of a single set of principles or an individual point of view. Depending on the situation and the factors involved, utilitarian cost–benefit approaches might form a dominant rationale for decision-making, while at other times questions related to fairness and individual rights may influence the path we follow. Yet, in environmental planning we bring to the decision an environmental perspective that places emphasis on maintaining balance between human need, environmental sustainability, and efficiency, along with an approach to planning firmly grounded in ecological principles.

The inclusion and consideration of ecological principles in planning and the decision-making process is the distinguishing characteristic that separates environmental planning from other expressions of the planning model. Ecological and environmental science is so central to this form of planning that it can be easily forgotten in professional practice. Nevertheless, to understand environmental planning theory, ecology, and the natural factors and principles that guide the environmental approach demands detailed examination and review.

Ecology's niche

The role of ecology in environmental planning and management is widely recognized. Examples of the work undertaken to demonstrate and refine the connection between ecology and planning include McHarg (1969), Park (1980), Steiner (1991), Westman (1985), and, most recently, Archibusi (1997). A common theme in this literature is the treatment of ecology not as science in its purest sense, but rather as a metaphor for synthesis. Ecology suggests a bringing together of elements into one combined system whose functioning takes on discernable patterns we recognize as the world we live in. This preoccupation with synthesis is not to discount the science of ecology, yet in planning the science is applied in a somewhat different way; as a language of design as well as one of description and explanation.

Ecology, by definition, is concerned with the study of the structure and function of organisms and their environment. As a science, ecology seeks to explain the interrelationships between living organisms and their environment and how they interact. This traditional view of ecology as the science of living things in relation to their environment provides the connection to the environmental planning problem and identifies ecology's major contribution to the management of the environment. When discussing the planning problem, we identify the need for planning as a response to the complexity and interrelatedness inherent to the landscape. The planning process was designed to manage that complexity and provide an avenue to direct understanding of how factors relate within the context of a given problem. Managing complexity and forming this fundamental understanding of relatedness does not occur in the absence of an organizing body of knowledge and theory. Ecology lends that knowledge and offers its theories to help support the environmental planning approach. Hence, through the lens of ecology we look at the planning area as a functioning "organism" and those things that comprise the planning area (people, houses, trees, water, etc.) all play a role.

As a science, ecology proceeds at three levels (Buchholz, 1993):

1 The individual organism.
2 The population identified as individuals of the same species.
3 The community composed of several populations.

At the level of the organism, the goal of ecology is to explain how individuals are affected by, and how they affect, their environment. At the population level ecology strives to describe the trend and fluctuations of a particular species and the factors that influence its presence or absence. When attention is directed at the community level, ecology seeks to define the structure and composition of communities and the processes that influence their living and nonliving elements. Translating these levels of analysis into the context of planning produces a model that can be used to examine the implications of an alternative and a strategy that can be used to frame our thinking with respect to

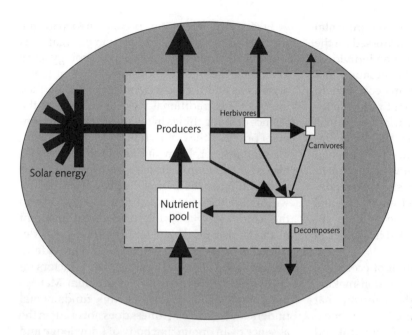

Fig. 2.1 Generalized character of terrestrial ecosystems.

the relationship between a plan and the levels of the environment that a plan may effect.

Perhaps the most important construct borrowed from ecology is the concept of an ecosystem. The ecosystem applies the logic of systems analysis to explain a device for organizing the complexities and connections that define the environment. To the ecologist, the ecosystem is the basic functional unit that incorporates organisms, populations, and communities together with the causal mechanisms that define the relationships, interdependencies, and pathways that direct energy flow and control the properties of the system. Taken into the realm of planning, the ecosystem concept can be used to form a root definition of the planning area, its living and nonliving elements, the relationships that connect them into a functional form, and the processes that govern their behavior. In ecology, it is common to consider the ecosystem as a complex of both biotic and abiotic components dependent either directly or indirectly upon the input of solar energy. The input of energy into the system produces a flow as it moves from one component of the system to the next. This flow establishes critical cycles of matter and energy that sustain ecosystem structure and complexity in a dynamic balance over time and space (Fig. 2.1).

Applying the ecosystem concept to the planning problem makes it possible to adapt a functional view of the landscape and structure elements of the problem into a representation that facilitates basic understanding of how things in the planning area fit together, what processes are actively influencing their structure, and how they change or modify over time. Using such a representation schema, we can explain the natural environment and characterize pathways that connect human processes to the natural system. The ecosystem framework further allows the planner to explore the dynamics of the environment using a process orientation that focuses attention on the flows or matter, energy, and resources through the system and permits definition of their direction and magnitude. This facility is perhaps the most significant quality of the ecosystem concept when used as an organizing framework in planning, simply because it enables the analysis of change to be grounded in an observable and measurable construct.

Using the ecosystem as an organizing structure with human activities embedded in it, and dependent upon it, the planner can trace the consequences of human actions on the model, describe the implications of human-induced change on the overall balance of nature, and assess the quality and stability of the environmental system. Functional stability introduces two valuable features of the ecosystem that help to frame the environmental planning approach: (1) self-maintenance and (2) self-regulation. Ecosystems maintain a dynamic equilibrium through a complement of self-maintaining and self-regulating strategies. As a consequence, ecosystems are constantly adapting and adjusting to natural fluctuations and perturbations that will influence community structure and composition. When human intervention is introduced, changes can be produced that create forms of stress that self-regulating and maintaining strategies cannot compensate for or adapt to. In these instances, the community can become threatened or critical functions can become "disorganized." This dynamic element of ecosystem development must be carefully understood by the planner, since it explains the principal source of conflict between the human-centered goals encouraging maximum productive use of the environment and the natural systems strategy of maximum support of complex ecological structures. Reconciling this conflict is the underlying purpose of environmental planning. However, to achieve reconciliation certain ecological principles must be incorporated into the foundations of the plan.

Guiding ecological principles

Environmental planning differs from other planning approaches by its inclusion of environmental factors into the decision-making process. However, planning with the environment, rather than against the environment, requires a sensitivity to the natural laws and principles that direct environmental functioning. It also requires a willingness to incorporate those laws and principles as a basis for design (McHarg, 1969). Planning in accordance with the principles and laws of ecology becomes one means to minimize the entropic ef-

fects of human development while encouraging greater balance between social progress and environmental process. The rationale for adopting these principles as planning guidelines is based on some very simple observations:

- Opportunities to develop and inhabit new areas have diminished.
- Development is increasingly directed toward areas or greater environmental sensitivity and constraint.
- Resources used to support the future will likely originate in environmentally sensitive areas where the costs and risks are greater.
- Postindustrial societies have become less willing to accept the social costs of growth and development.
- Greater consideration is being given to alternative and more sustainable forms of development at local, regional, and global scales.

Given these realities, connecting planning with the physical laws that govern the ecosystem further erodes the notion that human and environmental systems can be treated as non-interacting entities. By explaining these laws and principles in general terms it becomes possible to apply them and to demonstrate how they may exert a controlling influence on the scale and magnitude of human activities.

The laws and principles that direct environmental planning can be summarized and discussed under three major divisions: (1) primary physical laws, (2) laws of the biosphere, and (3) unifying ecological principles.

1 Primary physical laws

The relationship between energy and ecosystem maintenance can be explained in part by the fundamental laws of physics. Among the more germane to environmental planning are the law of conservation of matter, and the first and second laws of thermodynamics. The law of conservation of matter simply states that matter (mass) can neither be created nor destroyed, but merely changes from one form to another. To the environmental planner this suggests that materials are never really "produced" or "consumed." They change in form from raw materials and products to wastes and residuals without a change in quantity (Baldwin, 1985). Therefore, over time, in any stable system the amount of matter entering the system must equal the amount stored, plus the amount moving out. Therefore, those trucks that haul away the trash each week do not disappear once they round the corner. The matter they hold is simply being transferred elsewhere – nothing is ever really thrown away.

Energy interactions are governed by the laws of thermodynamics. The first law of thermodynamics, also referred to as the law of conservation of energy, states that energy can neither be created or destroyed, it changes from one form to another. As energy changes form and distribution, the quantity remains the same. Therefore, energy is neither produced nor consumed, it is simply converted from one form to another. The second law of thermodynamics introduces the principle of entropy. According to the second law of thermodynamics, within any closed system, the amount of energy in a form available to do work diminishes over time. This means that as energy changes form it becomes less useful, and less organized. This loss of available energy represents a reduction in a system's capacity to maintain "order" over time. To the planner, the law of entropy is important both with respect to natural and social systems (Baldwin, 1985). Because of the presence of entropy in order to sustain or enlarge any system, whether in the case of an organism or city, an expenditure of energy will be required. However, as energy changes form it become less organized, and because of entropy it takes more and more to yield less. Many of the management problems confronted in planning are a manifestation of entropy, whether those problems are expressed in a physical or social context.

2 Laws of the biosphere

The biosphere, or life-layer, describes a level of organization and an intricate set of relationships between its components and the physical environment they occupy. These relationships form a biotic pyramid whose shape represents the concentration of biomass and/or population at each level (Castillon, 1996). Using this pyramid concept it is possible to place elements of the ecosystem into a trophic structure that suggests where organisms relate in terms of the available energy and defines how they organize into food webs within the ecosystem. Five of the more central to the issues surrounding environmental planning include:

- **The law of production** – for the biosphere the law of production states that production must always equal or exceed consumption. Thus, at each level of the tropic pyramid, a carrying capacity can be defined which if exceeded will result in a reduction in population size or complexity.
- **Law of adaptation** – based on fundamental Darwinian principles, this law maintains that: (1) a species will produce more offspring that can survive, (2) these offspring possess the genetic traits of their parents, (3) these genetic differences in individuals of any species establish a competition for food and space, (4) those with genetic advantages survive the competition, and (5) the survivors pass the genetic advantage for survival to their offspring.
- **Law of fertility** – introduces the concept of nutrient cycling and maintains that nutrients must recycle to keep the environmental system functioning.
- **Law of succession** – with respect to biotic communities, this law states that there is a sequence of plant species that will occupy a recently altered or newly formed landscape.
- **Law of control** – introduces the idea that species have control mechanisms that govern population size and strategies for maintaining equilibrium given the nature of its habitat.

Taken together these laws or principles help to establish a better understanding of the underlying logic that guides environmental processes and basic biosphere functions. More importantly these law-like principles provide important support for environmental planning efforts by emphasizing equilibrium concepts and the importance of "balance" in maintaining a sustainable system.

3 General ecological principles

Because environmental planning strives to achieve a balance between development processes and environmental sustainability, development proposals and the motivations to transform land resources to more intensive forms of human use should be considered in relation to how "in tune" they are with respect to ecological realities of the planning area. Several ecological principles that govern the interaction of organisms and their environments exert a degree of control in establishing that relationship. These same principles can also be used to help direct environmental planning facilitate balance.

One ecological principle that carries important implications for the environmental planner is Ashby's Law of Requisite Variety (Baldwin, 1985; Margalef, 1970). Although this principle is more commonly expressed as the law of diversity and stability, the Law of Requisite Variety states that stable environments tend to develop diverse ecological communities over geologic time. Conversely, less diverse communities overall tend to be more vulnerable to environmental disruption. According to Baldwin (1985), the ramifications of this principle are fairly simple: "Don't put all your eggs in one basket." This fact holds for human systems as well as environmental given the general observation that social systems with greater economic and resource diversity tend to be more stable and tend to adapt more successfully to environmental, political, social, or economic changes. This principle also reminds the planner that diverse (high-variety) problems must be matched by a solution that equals the level of variety exhibited by the problem.

A second ecological principle that guides environmental planning has been referred to as the Brontosaurus Principle (Miller, 1998; Baldwin, 1985). Using an analogy with the ill-fated dinosaur considered the largest land animal and which became extinct approximately 75 million years ago, this principle asserts that bigger is better up to a point, whereafter size becomes a liability. The Brontosaurus Principle cements the notion that thresholds exist in the environment. When growth exceeds these thresholds it will be-

come increasingly difficult to manage the system involved. For example, when San Jose, California, was a city of 100,000 people it was comparatively easy to repair the streets, inspect the water lines, accommodate the needs of the population. Now as a city of nearly 800,000 encompassing large portions of the Santa Clara Valley, its not easy or cheap to tend to the day-to-day business of managing the city's infrastructure. In this context, the size of a community will constrain the ability of any unit to manage its size (complexity) and adapt to new conditions. We can further illustrate this point by considering the example of a small town that benefits from growth through the improvements made to public services, cultural amenities, and economic opportunities, to the point where that town is no longer small. At that larger size (both in terms of population and geographic scale) a "limit" (real or perceived) is reached where the benefits of growth have diminished and the city has to manage urban disamenities and a more costly, complex arrangement.

The notion that size imposes a constraint on the functioning of a system introduces the next and perhaps most influential ecological principle in environmental planning and decision-making, the carrying-capacity concept (Lein, 1993a). Carrying capacity can be defined as the maximum population that can be sustained by an ecosystem over time. In the context of environmental planning, carrying capacity describes the level of human activity that a region can sustain in perpetuity at an acceptable quality of life (Bishop et al., 1974).

The carrying-capacity concept was originally introduced in biology to explain the relationship between the resource base, the assimilative and restorative capacity of the environment, and the biotic potential of a species. The biotic potential of a species, describing the maximum rate of population growth that could be achieved given the number of females that reached and survived through their reproductive spans, is the controlling variable in this relationship. With adequate food supplies, living area, and the absence of disease and predation, biotic potential contributes to a growth in population that must be accommodated by the environmental system. Environmental resistance, however, regulates biotic potential by imposing limits on food supply and space, and other inhibiting factors such as predation and disease. Out of this interplay, carrying capacity emerges as the limit or level a species population size attains given the environmental resistance indigenous to its location.

In a planning or management context, carrying capacity is used to characterize the ability of a natural or human-made system to absorb population growth without significant degradation. Applying carrying-capacity concepts in planning requires careful treatment of four underlying assumptions (Lein, 1993). First, that there are limits to the amount of growth and development the natural environment can absorb without threatening human welfare through environmental degradation. Secondly, that critical population thresholds can be identified beyond which continued growth will trigger the deterioration of important natural resources. Thirdly, that the natural capacity of a resource to absorb growth is not fixed but can be altered by human intervention. Lastly, that how capacity limits are ultimately determined may involve an act of judgment. Although these assumptions make simple application of the concept difficult to uniformly apply, the concept of environmental carrying capacity has several important uses, including studies examining the effects of human actions on natural ecosystems, standard-setting for pollution control, and developing sustain-yield/renewable resource management programs. In each of these examples carrying capacity forms the basis for setting priorities and establishing levels of tolerance or thresholds in relation to critical environmental processes.

The concept of carrying capacity provides useful background to help understand a related unifying idea in the environmental sciences; the connection principle. The connection principle is based on the supposition that when examining the form and process of the environmental system, everything is somehow connected to everything else. While the scientific explanation supporting this assumption remains elusive, the philosophical meanings attached to the idea that all life is connected plays a central role in how the environ-

mental planner views the landscape. It also promotes a type of causal thinking where environmental relationships, actual or implied, become the framework for tracing cause and effect pathways. As Bush (2000) explains, every living thing survives by numerous and subtle relationships with all living things and the inanimate environment. When all living things are considered together, these connections appear as complex, interdependent, and self-regulating structures. Within these structures, any one form of life depends on the rest of the system to provide the conditions needed for its existence. While the earth has not always provided suitable environments for human habitation, over the millennia it was made hospitable by functioning ecosystems. The connection principle becomes one way to reconcile the theory that all things are intertwined in a complex web of hidden and subtle relationships that are presently beyond our comprehension (Baldwin, 1985). The value of this conceptualization of the environment to planning is that it forces the planner to examine cause and effect more carefully and to explore possible relationships that may not be obvious at first glance.

Although a more detailed discussion of central ecological concepts and their role in environmental planning could be undertaken, the physical laws and principles explored above provide a foundation to assist the planner in forming a better and more comprehensive view of the relationship between human systems and the natural environment. The principles reviewed above also remind the environmental planner that achieving a harmonious balance between human needs and the environment must be based on logic and a strong conceptual understanding of how the environment works.

The environmental planning process

The motivating purpose for environmental planning is to integrate environmental considerations into the planning process. With the inclusion of the environment, the planner is able to explore a wider range of alternative solutions and form a

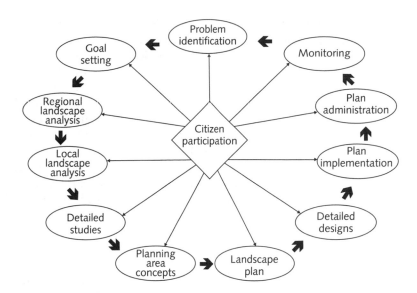

Fig. 2.2 The ecological planning model.

better understanding of their spatial implications. Through the integration of environmental variables along with socioeconomic criteria, a more comprehensible, efficient, and accurate information base for landscape decision-making is provided. This information feeds the planning process during the formulation, analysis, and selection of alternative solutions, and helps to determine which satisfy the established planning goals (Anderson, 1980). In this process the environmental planner introduces the "balancing" information that enables a broader interpretation of planning goals and an assurance that efforts to identify alternatives that promote sustainable development will not be compromised. In this sense, the environmental approach to planning asks us to look at the "other side of the coin" in order to see the complete picture.

Linking the environmental aspects of the planning problem into this process complements rather than replaces the stages of the traditional planning process. With the inclusion of the environment, the model can be reformulated as suggested in Fig. 2.2. Connecting to this information and to goals that are uniquely environmental, characterizes a new procedure for acquiring information and treating that information in a way that facilitates analysis and insight regarding the environmental controls that influence landscape

development. In this context, the environmental planning method is primarily a set of procedures for analyzing the biophysical and sociocultural systems of a place to reveal where specific development objectives may be practical with a minimum of environmental consequences (Steiner, 1991). This new procedure is defined by eleven interacting phases. As illustrated in Fig. 2.2, each explains a pathway for environmental information to focus decision-making, and together they describe a process where design plays a critical role in generating a successful outcome. This procedure, outlined by Steiner (1991), is an adaptation of the conventional planning process as defined by McDowell (1986), Moore (1988), and Stokes (1989), and builds on procedures developed originally for landscape planning (Marsh, 1983; Duchhart, 1989; Lahde, 1982).

The basic phases of this environmental planning model, based on Steiner (1991), can be summarized as follows:

1 Identification of planning problems and opportunities

The first step in the environmental planning method describes the exploration of issues that concern the interrelationships between the development process and the environment. This search basically requires consideration as to how development opportunities may conflict or adversely affect environmental resources and ecological functioning.

2 Establishing planning goals

Once specific issues have been identified, goals are established to address these problems. These goals provide the basis for the planning process and help crystallize an ideal future situation. Goal-setting is dependent on the cultural-political system and requires participation of all groups affected by a given issue.

3 Regional landscape analysis

This phase is unique to the environmental planning approach. Regional analysis describes the process of systematically characterizing the regional environment that constitutes the setting of the planning area. Generally, regional characterization focuses on the drainage basin or a related delineating feature that sets the region apart and enables detailed analysis. Within this regional setting, information is collected and a regional scale inventory of the natural and human factors relevant to the planning problem is produced. At this scale the collected information base is necessarily generalized and is used primarily to enable the planner to gain an "overview picture" of the region, its form, function, and situational characteristics, that allows important questions to be asked related to human/environmental relationships and permits simple "what if" scenarios to be explored that may suggest the need for more detailed studies. The purpose of conducting a regional analysis and inventory is to aid basic insight into how the regional system functions.

4 Local landscape analysis

Moving to a local-scale analysis concentrates on the collection of information regarding the appropriate physical, biological, and social factors that define the planning area. At this scale of analysis the goal is to obtain a detailed understanding of the natural processes and their relation to human plans and activities. Local analysis implies a much more in-depth inventory of critical natural factors and a classification of landscape characteristics that provides a systems view of human/environmental interaction. A key feature of local-level analysis is the addition of a sociocultural inventory that helps shape a cleared expression of human ecology within the planning area. Human ecology in this sense defines the manner by which people interact with each other and their environment, together with the relationships this interaction describes.

5 Detailed studies

Detailed studies connect the inventory and analysis of information to the problems and goals identified earlier in the process (Steiner, 1991). Although the term "study" can be a bit misleading in this context, this step in the environmental planning process describes the place where the information gathered during regional and local landscape inventory is subjected to specific analytical treatment. Anderson (1980) refers to this phase as landscape analysis modeling, since it represents the desire to form "models" of key environmental relationships that can be used to generate expressions of important environmental perceptions, values, and characteristics. Analytic models applied during this phase may include descriptive or predictive designs, or some combination of the two. In all cases, models are used primarily to evaluate specific conditions in relation to a goal in order to support:

1 specifying alternative programs or actions that might be chosen;
2 predicting the consequences of choosing each alternative;
3 scoring consequences or qualities according to a metric goal of achievement;
4 selecting the alternative that yields the highest score.

Common models applied to assist the environmental planner in these tasks include suitability models, separation models, vulnerability models, and attractiveness models.

6 Planning-area concepts

This step centers on developing concepts for the planning area. Planning concepts take the form of options derived from a conceptual model or scenario of how a goal may be achieved or a problem solved. These concepts typically express suitabilities or constraints produced through the combination of the information gathered during the inventory and analysis phases.

7 Landscape plan

Through the articulation of planning concepts, a series of preferred options are brought forward to serve as the foundation for the landscape (environmental) plan. The plan becomes a strategy to guide local development and offers flexible guidelines for decision-making. Topics given particular attention in this plan include considerations such as how best to conserve, rehabilitate, or develop the planning area. Unlike the comprehensive or general plan, which may have specific mandated requirements, emphasis here is on the landscape, giving greater importance to the natural and social factors that will direct the plan. As such, the landscape plan is designed to address the combined influences of land uses within the context of the total local/regional environment.

8 Citizen participation

The continuous involvement of the affected public remains the nucleus of the environmental planning model. Public participation occurs through a variety of educational and information dissemination programs. Beginning very early in the planning process, when issues are first being identified, public involvement introduces itself at each stage of the environmental planning process. This involvement is essential simply because it will help ensure that the goals emanating from the community are realized as objectives in the plan. With meaningful participation, important issues can be brought forward and resolved. If the public is included throughout the process, opposition to policy programs and recommendations will be limited, and the objectives of the plan will fit more closely with the desires of the community.

9 Detailed designs

The goals and objectives expressed in the landscape (environmental) plan will eventually influence the future spatial arrangement of land uses within the planning area. This design phase of the process offers decision-makers an opportunity to visualize the consequences of the policies and programs described in the plan, and to examine the geographic form and arrangement the plan will assume. Through careful assessment of various design alternatives, comparisons can be made regarding the short-term benefits of the plan with respect to long-term economic and ecological goals. Also, by rendering specific designs based on the landscape plan, the future distribution of land uses and the spatial organization of the environmental system can be critically reviewed before an irreversible commitment to development is made that may lead to adverse environmental changes.

10 Plan design and implementation

Making the plan work requires various strategies and procedures to realize the adopted goals and policies (Steiner, 1991). Several mechanisms common to planning include instruments such as:
    voluntary covenants,
    easements,
    land purchases,
    transfer of development rights,
    zoning,
    utility extension policies,
    performance standards.

11 Administration

Once the plan is adopted and implemented it must be administered to ensure that the goals and objectives established are achieved over the long term. Administration focuses on the role of planning commissions, citizen boards, review agencies, and other "overseeing" bodies that exist within the fabric of local government. The bodies

Table 2.1 Strategies and principles guiding environmental planning.

General planning principles:
Identify the planning process to be followed
Identify site-specific environmental goals and objectives
Evaluate new technology from ecological, natural resource, and social perspectives
Examine the conceptual and technical justification for a proposed project
Assess new project's environmental impact
Undertake environmental protection planning
Identify institutional capability for any recommendation
Predict for future scenarios the flexibility or reversibility of land-use decisions
Understand the compatibilities and incompatibilities among land uses
Communicate technical environmental information in an understandable form
Evaluate compliance with all regulations and applicable acts
Undertake an evaluation of risk and uncertainty
Incorporate environmental audits into project design
Evaluate assumptions

Natural science principles:
Understand historical ecosystem properties and trends
Undertake a systematic inventory of existing resources
Develop or adapt relevant ecosystem models
Predict thresholds, lags, feedbacks, and other constraint parameters
Identify natural processes and their significance
Identify key landscape indicators
Define pattern of opportunity and constraint
Identify unique geological and biological land units
Determine ecosystem stability–resiliency–diversity relationships
Determine carrying capacity and assimilative capacity limits
Identify significant transboundary ecosystem linkages
Monitor existing and built ecosystems
Design low-maintenance landscape systems

Social science principles:
Understand cultural linkages
Identify community and institutional values and individual perceptions and concerns
Design public participation approaches based on the level of interaction, representational needs and decision-making factors
Develop strategies to evaluate human values
Design outreach and educational programs to enhance public awareness

can be given the responsibility or charged with the duty to carry out or manage specific aspects of the plan, or to monitor and evaluate how well the plan is working.

The environmental planning process is also guided by a collection of technical principles that direct how the process unfolds and how the planner may exercise experience and judgment throughout each of the 11 phases. These technical principles are organized around three thematic areas – general planning, natural science, and social science (Dorney, 1989). These environmental planning guidelines have been arranged according to theme and are presented in Table 2.1.

Integrative environmental planning

How do we manage and direct the effects of change on the landscape? The answer to this question is not simple. To manage change we need to understand it and find ways to explain what it means in a way that makes environmental sense. In this section a "theory" is offered to help us place the connection between people and environmental change into something we can use as we observe and direct human use of the earth's surface. The explanation offered is highly integrative in nature and supports the idea that human use of and changes to the earth's surface are not something strange or unnatural. They are as much a part of this planet as the seasonal migration of birds and butterflies.

Integrated approaches to environmental planning and management have been widely advocated. Integrated planning explains the use of proactive or preventative measures that maintain the environment in good condition for a variety of long-range sustainable uses (Cairns, 1991). Given this definition, integrated environmental planning can be regarded as the coordinated control, direction, and guidance of all human activities within a specified environmental system to achieve and balance the broadest possible range of short- and long-term objectives. In this definition, integration implies synthesis and suggests that with environmental planning resting at the interface between the human/social system and physical/environmental system, a more realistic conceptualization of the planning problem begins to emerge.

Where are we going?

As a methodology, synthesis facilitates the combination of relatively simple parts into a more complex representation. From this representation, characteristics of the system as a whole can be deduced. Only recently has complexity been recognized to contribute important qualities to a system that cannot be predicted from a collection of its components. These attributes have been defined as emergent properties and provide the linking elements that connect the two systems together (Lein, 1989). For example, when one is considering the relationship or impact of land-use change on microclimate, the material structure and composition of the landscape can also be expressed as specific physical properties such as albedo or emissivity that can be measured and quantified. Early attempts at landscape synthesis focused on the interaction between population, economy, and environment, and on their territorial linkages. While such models may be criticized for producing a too highly generalized series of interactions, these approaches have merit for planning purposes, since they offer a general procedure for tracing the linkages between components of the landscape system. They also provide a useful methodology to explore the spatial aspects of interactions.

Synthesis also requires focusing on the interaction of diverse landscape features, drawing careful distinctions to guide assessment, and separate consideration of impact from that of consequence. Thus, through the introduction of synthesis in planning, a unity of process is realized that describes components of the planning area, including people, as elements of a larger system. Conceptually, environmental unity of this type enables the definition of interactions between components of the system that are not directly connected in operation yet exert an influence. For instance, it might be difficult to understand how economic growth might affect the flow of groundwater; the connection isn't obvious until we look for it. We can use the geosystem model to help gain insight into these larger connections. Here, the geosystem represents the combination of botanic, geologic, climatic, zoologic, and human

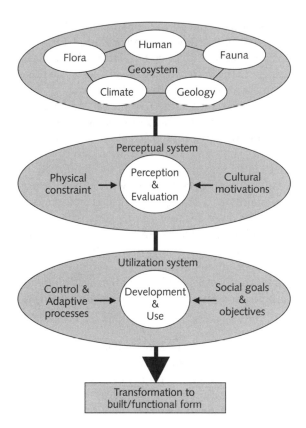

Fig. 2.3 Features of the land transformation process.

constituents that share the landscape. As expressed in relation to landscape development, people interact directly or indirectly with the remaining elements of the geosystem. Through this interaction an anthropogenic landscape is created (Fig. 2.3). This landscape describes an artificial arrangement of materials and energy that transfers its influence back through the geosystem. Anthropogenic landscape, therefore, explains a departure from natural surface form and explains specific surface alterations created by the development process in order to maximize utility from the landscape. These alterations also modify the morphological state (shape and form) of the surface, which changes the flow of materials and energy throughout the natural system. The resulting surface arrangement, its structure and origin, is therefore induced by human activity and becomes a visible product of planning. Just think about the number of times you've seen land being reshaped, raised, lowered to permit construction, the examples of drainage flows moved or altered, trees cut, soils replaced by concrete. Nature doesn't disappear in these areas, but it does look and respond differently, sometimes to our detriment.

However, until there are noticeable changes in the manner and intensity of use, most growth is too obscure to require attention. The isolated building lot or small modification of a road surface isn't going to cause significant change. To the planner this suggests that growth can be interpreted mainly as a change in the pattern of land use from less intensive toward more intensive uses of land where an increase in land-use intensity increases the productivity of land (Schafer, 1977). However, intensification of use can simply be a function of more people using the same land area as before, or it may involve other factors, such as:

- Higher capital investment in the same land area.
- Greater material investment.
- Increased productivity.
- The transfer of products or services.

Intensification also introduces competition. With land defined as a fixed and finite resource, users of land ultimately compete with one another in order to maximize the comparative advantage land enjoys due to its accessibility and resource quality. Emerging from this competition is an assumed rational allocation process that orders the distribution, type, and intensity of land use. At the local level, allocation is directed by the land market forces which determine the relative value of land and its preferred use, and the landscape begins to change.

How do we get there?

One of the better models that describes the spatial expression and dynamics of local land market processes was introduced by Alonso (1964). According to this conceptual model, two principal factors control the pace of development within the land-use system and direct its spatial form: (1) the economic rent of land and (2) its accessibility expressed as a function of distance to the focus of demand in the local economic system. Based on the interplay between these two factors, potential

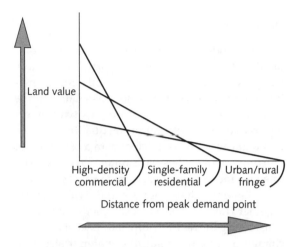

Fig. 2.4 Land-use allocation according to the bid–rent model.

users of land exercise a "bid" to locate at that spot based on (1) the anticipated economic rent (profit) to be derived from the intended use, and (2) the proximity that location offers to the sources of demand. The relationship between land cost (value), explained as a function of its accessibility, coupled with the pull of the market, creates a generalized pattern of land use (Fig. 2.4). The pattern suggested by this characterization suggests that at the surface there will be important variations in the intensity of use. Thus, based on this generalized description, there will also be a corresponding change in the degree of anthropogenic modification as the intensity or "impact" declines away from the locus of primary attraction. The spatial patterns that emerge, however, form only a partial explanation of anthropogenic landscape. The remaining portion of the equation directs attention to the physical consequence of this landscape as defined by its structural and textural properties.

Within the context of environmental planning, the built environment is more than a spatial representation of economic influences and trade-offs. This surface defines a form and texture that, while reflecting technological capabilities and human values, stands separate from these factors when viewed only in terms of its physical consequences. Through the eyes of the environmental planner it's not merely residential land, or commercial development, it's the material fabric those land-use types represent. We've all had the opportunity to look down at the landscape from the vantage point of an airplane window and we've marveled at how different these areas look: the mosaic or shapes, colors, textures; the curving streets and tree filled spaces of the residential area; the stark, wide, and smooth expanses of the shopping mall. These are real surfaces that behave in a physical sense and "do something" when environmental processes act on them.

That consequence becomes the responsibility of the environmental planner. However, the consequence of anthropogenic landscape develops as a function of time as this "designed" space interacts with the physical processes of the environment into which it has been placed. Placement is, therefore, a central feature of the problem and gives renewed emphasis to the plan and the spatial arrangement of future land uses. The precise nature of this placement is likely to be evolutionary as form adapts to a changing set of functional demands. Regardless, the interaction produced is directly attributable to the structural properties of designed space, whether expressed in the form of an agricultural field or a complex urban canyon.

Three underlying forces have been identified that shape urban morphology:

1 Forces that encourage the outward expansion of urban form.
2 Forces that influence the space and areal change in growth patterns.
3 Forces that control the intensity of land occupancy.

These forces include the contribution made by factors such as the pattern of land ownership, land speculation, personal income, land cost, and technological change in urban (built) form. Although these actors adequately explain the physical expansion and increased intensity of anthropogenic landscape, they do not address the concurrent structural or compositional transformations that occur at the surface.

Structural descriptions of landscape morphology must consider the physical composition of designed space and its variation over time. The fabric of this landscape becomes a mix of wood, brick, reinforced concrete, steel, and glass that represent the set of emergent properties that connect human activities to the larger environmental system. Therefore, the planner's allocation of functions to a specific point in the landscape also becomes an indirect decision regarding the new properties that will take hold on the landscape and how those properties react with the natural forces that drive the landscape system. Will the new conditions introduced be beneficial to the functioning of the environment, or will they introduce a stress that will eventually undermine its sustainability?

Building toward sustainability

As noted in the previous section, the introduction of built form presents a new morphological state capable of moving the environmental system to a new equilibrium, bringing into question the issue of sustainability. As illustrated in Fig. 2.5, the predominant threat to sustainability emanates from the stress imposed on the natural system when land is changed from its natural state to some other form. According to Fig. 2.5, the environment in its natural condition can be explained by a use potential that translates into an expression of the landscape's worth, and a use value that describes the ability of the landscape to satisfy one or more societal needs. If a benefit can be identified then the development of that area may be encouraged. The process of development in this context introduces human impact; which for the most part involves the removal, replacement, redistribution, and redefinition of the geosystem. Through this process, the surface is transformed into an explicit functional state. To the environmental planner, the surface now contains distinct functional attributes relating to its use (building, roads, utility lines), a morphological characteristic reflecting its physical structure (a shape), and a set of inherent qualities that define the invariable or stable features not affected by human activity (environmental conditions that have not been modified).

The presence of this new "anthropogenic" state produces real and potential effects as the environ-

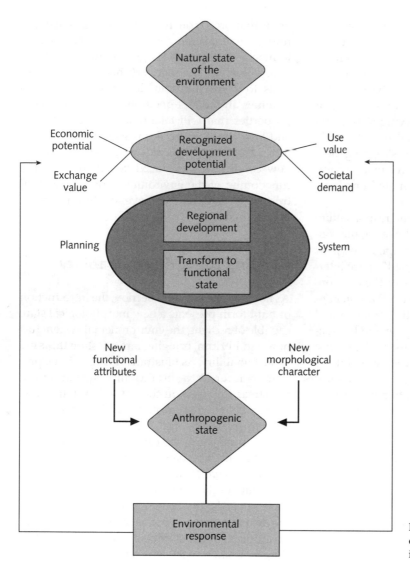

Fig. 2.5 Defining environment/development interactions.

ment responds to change. These feedback effects flow directly into the socio-economic regulators propelling development. The environmental response, as depicted in Fig. 2.5, may initially explain qualitative changes in the structure of the geosystem that may eventually alter its performance. At this point, the capacity of the landscape to satisfy societal demands may be compromised, and the benefits derived from use of the future economic potential of the landscape may be degraded as more resources must flow into management in order to sustain the pattern.

The consequence of human development and intervention as implied by this simple model is similar to the introduction of a stress into any natural system. Stress, in this context, identifies a form of interference with the expected condition of a system (Lugo, 1978). Of course, the effects of stress are most dramatically observed after critical thresholds of tolerance are exceeded and a strain of anomalous response in the system is produced. When this occurs a departure from the expected condition results and a new state develops to which all other elements in the system must adjust. If we abstract the geosystem according to the relation

$$Y = f(x, p, t)$$

The system (Y) is composed of a vector of state variables (x) and a vector of parameters (p) with (t) representing time. By invoking an arbitrary function to represent the state of the system,

$$\beta = f(x_1 * x_2),$$

a strain (s) can be explained. This strain develops out of the deviation in the state β and can be denoted as β^*, such that

$$s = /\beta - \beta^*/$$

where,

$$\beta^* = f(x, \phi),$$

with ϕ representing the change in one of the original state variables. A sustainable system, therefore, defines the condition where $/\beta - \beta^*/ = \phi$ and a strain is not produced that results in a system response.

Realizing the condition expressed above where any strain is kept well below threshold levels depends greatly on the approaches available to encourage environmentally sound patterns of development (Naess, 1993). A sustainable pattern is not possible unless development proceeds in accordance with criteria for achieving sustainability. Opinions regarding what is an environmentally sound form of urban development differ. In a review of the topic by Naess (1993), several major themes can be identified to help focus the concept of sustainable development. For example, transportation planners tend to recommend concentrated development patterns (Owen, 1986; Newman & Kenworthy, 1989), while those working in residential design give preference to urban expansion over density increases (Attwell, 1991). Finally, others argue that regional satellite developments are an optimal strategy for combining conservation goals with rural preservations (Naess, 1993).

Recognizing options

In general, by concentrating the development pattern, confining new construction to areas where technical encroachment on the natural system has previously taken place, and utilizing each building site efficiently, the preservation and maintenance of the environmental system becomes possible. Development must therefore respond to a series of goals that together realize sustainable patterns of growth and change. These may include attempts to

1 Minimize energy consumption.
2 Preserve biological resources.
3 Minimize costs.
4 Minimize local noise and pollution.
5 Provide opportunities for outdoor recreation.
6 Preserve landscapes and cultural values.
7 Contribute to the realization of social goals.

It must be recognized that while these goals will do much to improve the present situation, true sustainability may never be achievable (Beatley, 1995; Rees & Wackernagel, 1996).

The major limitation of the concept of sustainability is that it has become a "buzzword" in current planning and resource management discourse (Collicott & Mumford, 1997). When applied, its anthropocentric focus is not always useful within a strict ecological framework. Thus, sustainability should be replaced with the more exacting concept of ecological sustainability. Ecological sustainability takes into consideration more concretely a wider concern for ecosystem health and links to the objects of environmental planning. An attempt to place sustainability concepts into a more ecological context was introduced by Lawrence (1997). Although this outline directs attention toward the environmental impact assessment process, its features are useful and relevant to the general concerns of the environmental planner. The attraction to this framework are the instruments introduced to meet sustainable development goals (Table 2.2). Using these instruments, a conceptual foundation is presented that connects sustainability to the larger environmental issues (Lawrence, 1997). The model suggested by Lawrence illustrates how sustainable planning can take place at the conceptual, regulatory, and applied levels. If sustainability concepts are integrated into the planning

Table 2.2 Fundamental goals directing sustainable development.

Approach problems from a sustainability systems perspective
Adopt a long-term view of human–environmental conditions
Strive to span jurisdictional, disciplinary, professional, and stakeholder boundaries
Ensure that values and value differences are made explicit
Keep options open
Be sensitive to the consequences of being wrong
Ensure that the means to achieve sustainability are sustainable
Design approaches to suit the context of community need and aspirations
Ensure a full accounting of social and environmental costs
View global environmental management as a shared responsibility of all

process, changes should be produced in the ways in which environments are managed. The main obstacles frustrating this ideal are those of implementation and the larger problem of generating an understandable method to express the concept in substantive, quantifiable terms.

Forming a useful expression

Recently Rees and Wackernagel (1996) have suggested an approach to quantifying sustainability they term "our ecological footprint." Adapting this approach to planning may produce a better expression of the ecological impact of the urban "built" environment. Their approach is based on the observation that urban development represents a human ecological transformation. Therefore, the shift in human spatial and material relationships with the rest of the environment is the critical link to sustainability. Using the definition of environmental carrying capacity as the maximum persistently supportable load, sustainability can be redefined with carrying capacity as the operator governing its expression. According to this logic, no development path is sustainable if it depends on the continuous depletion of productive capital. Thus, to foster sustainable development, a critical amount of such capital must be conserved intact and in place. This will ensure that the ecosystems upon which humans depend remain capable of continuous self-organization and production (Rees & Wackernagel, 1996).

Realizing this goal depends on forming better definitions of human carrying capacity and the maximum entropic load that can safely be imposed on the environment (Catton, 1996; Rees, 1996). The "footprint" concept recognizes that as a consequence of the large increase in per capita energy and material consumption made possible by technology and global-scale trade dependencies, the ecological locations of high-density human settlements no longer coincide with their absolute geographic locations. Our feet have simply outgrown our shoes. Because cities "appropriate" the ecological output and life support functions of distant regions, it is also critical to recognize that no urban region can achieve sustainability on its own (Rees & Wackernagel, 1996).

Determining the ecological footprint of a population level residing in the urban landscape facilitates this understanding by:

1 Estimating the annual per capita consumption of major consumptive items from aggregate regional data and dividing total consumption by population size.
2 Estimating the land area appropriated per capita for the production of each consumptive item by dividing average annual consumption of that item by its average annual productivity or yield.
3 Compiling the total average per capita ecological footprint (EF) by summing all the ecosystem areas appropriated by an individual.
4 Obtaining the ecological footprint (E_{fp}) of the planning area population by multiplying the average per capita footprint by population size ($E_{fp} = N \times EF$).

More complete details of this procedure can be found in Rees and Wackernagel (1994), Wackernagel and Rees (1995), and Rees (1996).

Although ecological footprint analysis is not a form of dynamic modeling and has no predictive capabilities, the method acts as an ecological camera that produces a snapshot of a population's current demand on nature. For the planner, such a snapshot can be a useful means to compare development proposals and evaluate growth policies with reference to the environmental support that will be required to accommodate a given alternative. However, perhaps the greatest obstacle to overcome when approaching the topic of sustainability is the negative expectations people have about what a sustainable society will look like (Beatley, 1995). Because a sustainable society would likely shift focus away from materialistic quantities and toward more abstract measures of quality, it conveys the image of a spartan and materially backward way of life. Because planners are visionaries, we must learn to describe sustainable futures in ways that effectively communicate alternative visions. Much of this begins with the plan, what it says, and how it communicates a future to those who stand to be affected by it. In the next chapter we will explore the nature of this plan, how it takes shape, and the important considerations that guide its formulation.

Summary

The nature of environmental planning served as the subject matter for this chapter. Set apart from other branches of the planning discipline, environmental planning was defined as the one approach to human landscape development uniquely devoted to achieving balance between human needs and environmental quality. From its philosophical roots and the ethical responsibilities that direct us at present, environmental planning, with its focus on integrating environmental ideals and information into the planning process, becomes an important step to ensure long-term sustainability. Achieving sustainable goals requires an understanding of how to plan environmentally. This chapter reviewed the ecological principles that guide environmental planning and pave the way for an integrative view of the planning problem based on synthesis and a conceptualization of human development as a physical feature of the landscape. By considering this perspective, and through the environmental planning process outlined, more sustainable landscape patterns can be realized.

Focusing questions

What do terms such as compatibility, suitability, susceptibility, and sustainability communicate when placed into the environmental planning process?

To what extent can natural laws provide guidance in the design of environmental plans and policies?

Explain the value of looking at human development through the lens of anthropogenic landscape.

How can the conceptualization of stress in a system be used to structure environmental planning problems?

CHAPTER 3

Making Plans

The planning process described in Chapter 1 and the principles that define environmental planning discussed in Chapter 2 ultimately become embodied in a plan. This plan expresses the goals and objectives of a society that will guide the allocation of functions within the land-use system to produce a desired future state. How a plan takes shape, its design, and the influences that direct its formulation are the major themes of this chapter. Here we will examine the technical aspects of the plan as a decision-making document, together with the philosophical foundation that anchors the ideological features of design to specific physical outcomes.

A conceptual view

A plan can be thought of as a blueprint for the future. It presents general goals and objectives of the community and blends them with specific policy recommendations developed with the single purpose of moving the community closer to some desired future. Earlier, we spoke in general terms about the nature of goals and objectives within the context of the planning process. In that discussion these ideas were vague and suggested that something meaningful drives planning. We now want to take that background information and expand on it and examine what actually goes into a plan and what it means.

From a purely pragmatic point of view, a plan is an official public document adopted by a local government as a policy guide to decisions it will make regarding the physical development of the community. Whether it is termed a general plan, comprehensive plan, community plan, or something similar, the plan indicates in direct language how government leaders want the community to develop over a broadly defined time horizon. Typically, time is expressed in a plan using increments of 10 to 25 years and sometimes longer. However, in all cases, the expectation implied in a plan is that the goals expressed within it will be realized gradually over this time horizon. Precisely how well those goals were realized over the expressed time horizon has received comparatively little attention. Yet auditing and monitoring the plan is an important part of ensuring that its role as a policy tool remains constant and that decision-makers follow the specified policy recommendations consistently over the time horizon.

It has often been stated that the essential characteristics of a plan are that it is comprehensive, general, and long range. Although these words are easy to use, their meaning frequently blurs. The term "comprehensive" suggests that, to be useful, a plan should encompass all geographical parts of the community and all functional elements that influence physical development. If the plan is not complete in its characterization of the planning area, and if certain critical features are omitted from consideration, it will not provide the guidance or detail needed to direct change. A

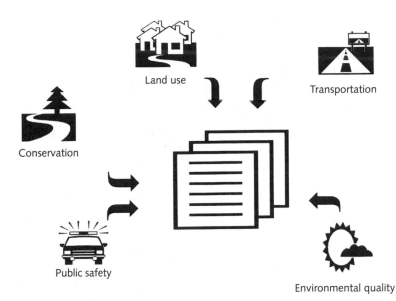

Fig. 3.1 Typical elements
incorporated into plans.

Land use

Transportation

Conservation

Public safety

Environmental quality

partial plan has limited value given the connectedness that defines the planning area, its environment, and the complexities that describe human interaction within this mix. Similarly, the term "general" implies that, to be effective, a plan should summarize policies and proposals, but not provide specific locations or detailed regulations. If a plan introduces too specific a design it leaves little room to adapt to changes that may result over the time horizon. By maintaining a more generalized posture, recommendations can suggest specific changes that policy-makers can enforce through existing laws and regulations, or identify gaps where new laws or regulations may be required. Finally, the concept of a long range directs the plan and all involved in its creation to look beyond the foreground of pressing current issues and consider instead the problems and possibilities 10, 20, 30 years into the future. Long-range thinking directs focus on proactive decision-making. Although not typically a feature of a culture grounded in the immediate satisfaction of wants and needs, a plan that does not assume a proactive stance provides little guidance to those who must decide on the allocation and distribution of scarce resources, or to those concerned with management of a sustainable land-use/environmental system. Given these essential qualities,

what exactly goes into this document and how do all the parts fit together to produce a workable "blueprint"?

While plans will vary in content and format, certain elements are common and form the salient characteristics of a plan and its focus. (See Fig. 3.1.) These fundamental topics of interest and concern include:

- **Land use** – describing the current characteristics of the land-use system, future conditions that may arise, together with policies and programs directed at specific land-use issues or development goals.
- **Transportation and circulation** – explaining the existing road network, traffic conditions, and anticipated future conditions with policies and programs designed to address specific transportation needs and goals.
- **Public safety** – characterizing natural and human-made hazards including geology, floods, hazardous materials, wildfires, and other potential sources of risk within the planning area, along with policies and programs designed to reduce human injury, loss of life, property damage, and social-economic dislocations due to these events.
- **Conservation** – describes existing natural resources within the planning area and

presents goals and policies designed to enhance the conservation and management of natural resources and open space, the preservation and production of resources, the promotion of outdoor recreation, and the protection of public health and safety.

- **Environmental quality** – discussing pollution factors and concerns such as those affecting noise, air, water, and soil with specific reference to existing pollution levels, comparison to standards, identification of sensitive receptors, and goals, policies, and programs targeted at major environmental quality issues.

Other elements may also be found in a plan, including sections devoted to the analysis and assessment of housing, education, or public facilities (Kelly & Becker, 2000). Still, in other cases, specific elements may be mandated by statutory regulations specific to a given governmental authority.

Regardless of contents, certain analytical inputs are common to all elements that comprise a plan, and these guide its physical development. These are the skills of the planner and the tasks that are performed as data is transformed into information and placed into the plan: (1) description and documentation, (2) definition, (3) projection, and (4) prescription. Taken together they explain the intellectual skills brought to the plan and the principal methods used to communicate its features to decision-makers.

1 Description and documentation

Because a plan summarizes existing conditions and provides critical background information to help improve basic understanding of the planning area, description is an essential feature of the plan. Description, however, does not occur in a vacuum. Rather, description implies the careful collection and selection of data that will effectively characterize the important features, qualities, and quantities that will be discussed in the plan. As an intellectual activity, description involves several interrelated purposes. At the most basic level, description enables the decision-maker to "see" the characteristics of the objects and features under

consideration. For example, in a certain midwestern community in the US, it was decided that the time had come to update the comprehensive plan. The last major revision to the plan was done in the 1970s, and civic leaders recognized the need to begin again. Since so much time had elapsed since the last plan was made, planners needed to begin from square one and collect and summarize the general social, economic, and environmental characteristics of the community, so they could assess the current state of the community and begin to piece together an explanation how the city has changed since the last plan.

Therefore, to be useful, description must therefore provide insight into the patterns and processes that explain the arrangement of objects and features that constitute the planning area. In this sense, description should provide important cues as to

- Why the pattern is what it is.
- How that pattern came to be.
- What factors influence its disposition.
- What makes this pattern significant.

Although description may not demand a detailed geography of every topical area that might define some aspect of the planning area, it should be systematic, organized, and allow someone not familiar with the planning area to gain a reasonable understanding of the subject matter. To illustrate the role of description, consider the contents of a conservation element that might be found in a general plan. This element might begin with a review and inventory of existing resources. Within this section the natural resources, including various categories of land cover, native plants, native fauna, historic sites, and amenities, would be subjected to detailed examination and representation in the form of maps and diagrams. For each factor, description would concentrate on the location of these resources, the site and situational factors that characterize their location, and their physical presence in the landscape. Following such a treatment, the dominating features that distinguish these areas could be discussed, which might include detailed explanations concerning specific plant or animal species and maps that delineated these areas and placed them into a geographical context with respect to the total planning area. In

Table 3.1 Common information sources.

United States Geologic Survey
US Department of Agriculture Soil Conservation Service
National Oceanic and Atmospheric Administration
US Army Corps of Engineers
State Departments of Natural Resources
Water districts
Utility districts
Historical societies
Universities and colleges

addition, factors such as soils, agricultural lands, groundwater recharge zones, and recreation areas may also be included in a description of the resource base. To help frame these descriptions, statistical data is frequently used to establish magnitudes and summarize quantities to help the reader gain an appreciation of scale and importance. Since the main purpose of description is to give the reader background, care must be given to ensure that only those salient factors are included in the discussion. This requires the exercise of judgment in selecting those factors that give the subject definition and significance (Table 3.1).

Description also serves another purpose, that of documentation. It is not uncommon or surprising that decision-makers or members of the public will be ignorant with respect to the detailed aspects of their environment. No person should be assumed to have complete knowledge of the groundwater recharge zones, housing conditions, demographic characteristics, or any other combination of variables that define the environmental complex that is the planning area. Describing these features, depicting their geographic location and extent, or providing an inventory or accounting of their nature and distribution, documents and records their existence. This record contributes to a better understanding of their significance and provides critical baseline data against which change may be assessed.

2 Definition

Recognizing that a plan represents a policy instrument, it will necessarily require the use of language and concepts that can be inexact or ambiguous in meaning. While it may not be the intention of the planner to obfuscate, the potential to confuse those reviewing the plan always exists. Definition plays a dual role in this regard. First, definition does much to provide detailed explanations of terms and concepts used in the plan whose meaning may be unclear or unfamiliar to the audience. Secondly, definition helps fulfill the obligation of disclosure, ensuring that critical ideas, issues, or information are communicated clearly and effectively. Because a plan is written for all members of the community, the document should be written in language that can be read by a lay-audience. There is no single more frustrating experience for the planner than to be making a public presentation of a plan and be asked to explain what terms and details mean because they were aimed over the "heads" of the audience; and nothing calls the planning process into doubt more than "double-speak" that erodes public trust and confidence. Technical terms must be defined throughout the text and detailed technical data must be referenced in supporting documents. The physical organization of the plan is therefore critical. In purely mechanical terms, a plan will be divided into chapters (sections) including an introduction, and while a plan is not read cover to cover in the same fashion as one reads a novel, the arrangement of those sections should be logical. An example of a generic table of contents that could be found in a typical general plan is shown in Table 3.2. Although this instance is hypothetical, each chapter will contain two principal sections, one that describes the present and future conditions of the topic with respect to the planner area, the other discussing community goals, policies, and implementation programs.

Definition also carries another important interpretation within the context of developing the plan. This aspect is perhaps more critical than the technical questions surrounding the mechanics of its contents. Here, definition speaks to the issue of defining the planning problem and the relationships that characterize the variables that control or influence the problem. Specifying and defining the variables and their relationships that will be discussed in the plan draw on the knowledge, experience, and judgment of the planner. Through

Table 3.2 Basic elements of a plan.

Introduction
  Purpose and nature of the plan
  Role of the planning process in local government
  Relationship of the plan to other plans

Background
  Historical background of development in the community
  Current conditions and trends
  • the built environment
  • the natural environment
  • the economic environment
  • the social environment
  Current and emerging issues that carry long-term
  implications

Assumptions
  Assumed effects of external forces on the future of the
  local community
  • physical developments
  • social developments
  • economic development
  • political developments
  Local policies, values, and actions that will affect
  development
  Regional goals and issues
  Forecasts of regional and local growth

Overview of the plan
  Community goals and objectives
  Basic community design concepts
  Major design proposals

Major implementation strategies

Planning area maps and diagrams

Planning elements
  Land use
  Circulation
  Community facilities
  Utilities
  Transportation
  Housing
  Public safety
  Natural resources
  Conservation and open space
  Natural hazards
  Environmental quality
  Noise
  Growth management

this process of specification several important factors must be considered:

  • Definition of attributes – explaining what the variables are and how/why they relate to the problem.

  • Factors that affect attributes – explaining the types of activity or process that influence the status or condition of a variable.
  • Qualities or quantities to measure – explaining the characteristics of an attribute that lend themselves to measurement.
  • Methods of measurement – explaining the specific methods available to quantify the attributes and express their salient features.

Identifying the problem and the attributes that describe it provides the logical connection between specific planning issues and the goals and policies that will follow to address them.

3 Projection

As a document designed to guide the long-range management of a community's land and natural resources, a plan employs numerous methods of forecasting to complete and evaluate various problem scenarios. For example, trends in economic growth and job creation may be used to develop future population growth. Those projections, in turn, may be used to estimate anticipated demand for residential land use. Estimated changes in residential land can be used to project changes in open space, zoning, resource use, utility demand, and energy consumption. In each case, effects can be evaluated and policies or programs can be developed to address each new situation in relation to the goals expressed in the plan. For example, the Baltimore Ecosystem Study has made extensive use of projection methods to understand trends in development around the Chesapeake Bay. Here, projection methods are being used in order to form better policies aimed at controlling environmental impacts associated with intensified human use of the region. Sources examining the role of projection in decision-making and the use of models in planning include Lein (1997), Klosterman et al. (1993), and Gordon (1985).

A variety of projection techniques are used in planning. In the majority of cases, projection involves the application of a model. This model typically falls into one of three descriptive categories: (1) digital process models, (2) spreadsheet models, (3) general-purpose simulators (Lein, 1997).

1. **Digital process models** describe computer simulation models developed to simulate key socioeconomic, environmental, or physical processes. Models of this class describe specialized computer programs designed to function as stand-alone packages that require data and parameters placed into formatted input files. Such models generate their own output, be it a projection of population, traffic volume, air pollution concentration, or level of nonbasic employment demand. Current use of any computer model requires an understanding of the problem under investigation, together with the specific information that controls operation of the model (Lein, 1997). Selecting the appropriate digital model as an aid to planning occurs after the information to be gained from the model is understood and the data required to drive the model is collected.

2. **Spreadsheet models** define programs that have been developed using common spreadsheet or data management software packages. Models of this type store data as a two-dimensional table, permit calculations with the data, and instantly display results in a variety of graphic formats. Because of their widespread use in budgeting and administration, spreadsheets are common and are an attractive alternative for data analysis and modeling (Hardisty et al., 1993; Cartwright, 1993; Klosterman et al., 1993). Sophisticated spreadsheet programs offer an array of built-in functions that greatly enhance computation, database management, file import and export, and display.

Although these systems were designed primarily for financial analysis, they have evolved into programming environments and offer several features that can enhance the use of projection in planning analysis (Lein, 1997):

- Relative ease of programming.
- Comparative ease of modification.
- A transparent design.
- A functionality that provides power and flexibility.
- A capability to generate an assortment of graphics.

The major disadvantages associated with spreadsheet models are speed and processing capabilities. In general, models written using spreadsheets tend to be slower and less elegant when compared to those developed using a standard programming language. Furthermore, because of their spreadsheet format, iterative processing is more difficult to implement, which can limit their application to dynamic systems. These drawbacks notwithstanding, numerous models applicable to a range of planning problems have been introduced (Klosterman et al., 1993).

3. **General purpose simulators** describe a family of computer languages developed specifically to support modeling efforts. The main advantage of this approach to projection is that these languages provide a simple syntax for developing a model that improves the representation of process and the characterization of complex systems. Several of these features have been summarized by Lein (1997) and include:

- Controlling events.
- Collecting and representing data.
- Generating random variables.
- Managing simulation time.

Of the general-purpose simulators available to the planner, those with the capacity to model continuous systems of events are of particular interest. An excellent demonstration of such models can be found in Hannon and Ruth (1994). Using a graphic-based simulation language, a range of dynamic problems can be examined, from pollution and ecological process to economic modeling.

A variety of topics may be subject to some form of projection into the future. Typical features of the planning problem subject to projection include land use, transportation, population, air and water quality, noise levels, and employment. A more comprehensive listing of landscape variables that can be used to form projections is presented in Table 3.3.

4. **Prescription.** In planning, to prescribe means to direct the use of land and other resources as a remedy for specific social, economic, or environmental problems. As a sequence of actions that become realized in the plan, prescription defines a multistage process that involves:

a) Exploring the problem and forming a basic understanding of the relevant objectives and values.

b) Producing a set of alternative choices.

Table 3.3 Commonly projected variables in plans.

Human/economic:
Demographic factors
Land-use change
Employment
Income characteristics
Transportation flows
Solid waste generation
Housing factors
Noise conditions
Energy demand
Spatial interaction
Demand factors

Physical/environmental:
Air quality
Water quantity/quality
Groundwater processes
Flood Processes
Land Cover Change
Habitat Characteristics
Erosion/Sedimentation
Hydrologic Systems
Food Chain/Food Web Dynamics
Pollutant Fate and Transport

c) Identifying the adverse and beneficial properties of the alternatives.
d) Evaluating alternatives.
e) Recommending the "best" alternative as the optimal solution.

The solution is generally some form of regulatory policy or program that may be expressed as either an objective that can be maximized or a constraint that can be minimized. Given the reality within which planning operates, each objective and constraint carries a political weight, whether stated explicitly or not, and that weight can influence how the "best" alternative is defined.

Directing the use of land and other resources also suggests that prescription is a form of analysis embedded in a decision-making process. Prescriptive analysis may be highly formalized, or it may remain an informal procedure. Formal methods of prescription identify very structured approaches to problem-solving and may involve the use of well-understood procedures for simplifying the process of selecting the optimal alternative. Informal methods recognize that plans grow out of political and professional deliberation, negotiation, and bargaining. Optimal solutions, in this context, are defined not in terms of a single quality or quantity, such as might be produced from a formal methodology; but rather on the basis of professional judgment, public opinion, or a multidimensional political perspective. With either approach, the goal is the same: to narrow down the possibilities and identify a workable solution.

Plan formulation

A plan is designed to fulfill three important purposes:
1 To facilitate the process of making policy.
2 To communicate that policy to all interested and affected parties.
3 To assist in the implementation of policy.

The process of formulating this plan can be divided into stages. Each stage produces a particular type of plan or specific element. The sequence of stages and the products generated by each suggest a progression from ends to means, as well as from general policy statements to specific programs (Kaiser et al., 1995). The basic stages that outline the process of plan formulation are given in Table 3.4. As illustrated in Table 3.4, the first stage in formulating a plan begins by developing a comprehensive understanding of existing and emerging environmental conditions. Drawn from this initial description are a set of implications that identify problems or concerns around which goals and objectives are created. For instance, in our earlier example of the mid-western city, let's suppose that during the description and documentation process it was noted that new development in the region has occurred in areas where soils have a high erosion potential. Since this might be a contributing factor to the changes in stream water quality that have also been documented, goals to address both the development issue and water-quality changes will be likely components of the plan. Stage two of this process extends the goal-setting and problem formulation stage by adding explicit prescriptive studies that: (1) identify the present and future demand for land resources; (2) specify areas of critical concern, such as lands to preserve to enhance natural processes, lands to

Table 3.4 Stages in the plan formulation process.

Stage 1: **Direction-setting**	Describing existing and emerging conditions and causes Setting Goals Formulation of general policies
Stage 2: Land classification	Analyzing basic land demand and supply Designating areas for natural processes Designating areas for urban use Designating areas for agricultural production
Stage 3: **Land-use design**	Analyzing detailed land demand and supply Designating locations for employment and commercial centers Arranging residential communities and facilities Designating locations for infrastructure and community facilities
Stage 4: Development management	Analyzing implementation factors Setting procedural goals Specifying components of the plan Specifying standards and procedures

protect for agricultural production; and (3) developable areas suitable for future use.

Several important tasks are associated with this stage of plan formulation. Each of these tasks becomes the subject of systematic analysis and defines the core information presented in the plan. These tasks include:

1 Developing locational standards to guide land-use allocation.
2 Deriving a geographic expression of suitability to guide land-use placement.
3 Determining the amount of land that will be needed to meet anticipated future demand.
4 Defining carrying-capacity levels given to estimated available land.
5 Exploring alternative location arrangements and designs for future development activity.

The third stage in this process builds on the foundation completed in stages one and two. Here, attention is focused on the optimal geographic allocation of functions to specific loca-

tions in the landscape. At this stage of plan formulation consideration is given to the holistic aspects of design and the future land-use system's relationship to critical environmental parameters. By considering the future arrangement of housing, commercial uses, industrial facilities, parks, open space, and infrastructure with respect to maintaining environmental quality, the plan can become a reasonable "blueprint" that guides development with a minimum of environmental disruption.

The final stage of plan formulation considers the vexing problem of how this optimal future arrangement can be achieved. Emphasis during this stage is given to the design and implementation of programs to institute development regulations, capital improvements, and incentives that local government can employ to direct land use and environmental change. A selection of instruments that government may call upon to direct and manage changes in the land-use system are provided in Table 3.5. Within the context of the plan, this stage produces a sequence of recommendations that can be adopted and implemented over the time horizon of the plan.

To illustrate this aspect of the process and its language, consider the following community goal that might be expressed in a hypothetical plan:

To preserve the natural and "manmade" resources of the planning area, including plant and animal habitats, water courses, and historic structures.

To meet this goal a series of policy statements lined to specific programs are offered. One example might include:

Policy 1: Preserve those natural wildlife habitats which support rare and endangered species of plants and animals where appropriate.

Making this policy reality requires connecting it to some type of action-forcing mechanism, such as:

Program 1.1: Restrict development to one single family home on existing lots of record within

Table 3.5 Government tools to implement and direct plans.

Corporate powers – directing land acquisition and development
  Construction of streets, roads, and water and sewage
    treatment facilites
  Acquisition and development of parks
  Acquisition of sites for low and moderate income housing
  Purchase of development rights and scenic easements
  Creation of development corporations

Police powers – regulatory action
  Specific plans
  Zoning:
    • open-space zoning
    • environmental-hazard zoning
    • inclusionary zoning
    • planned unit development zoning
  Subdivision regulations
  Park dedication requirements
  School dedication requirements
  Review and regulation of public works
  Housing and building regulation
  Code enforcement
  Environmental review procedures
  Design review

Other options
  Redevelopment
  Intergovernmental cooperation
  Public information
  Data management
  Monitoring
  Cooperative arrangements with private sector

nondevelopment portions of designated habitat areas.

Program 1.2: Designate the majority of upland areas Public Health and Safety districts to protect wildlife habitat.

In this example the two programs direct the use of land in specific ways, and designate land areas that restrict use or limit the use of land in areas where analysis has shown them to have important habitat functions. Using these programs to direct future land use is seen as one way to meet the goal of preserving habitat within the planning area.

Next, let's consider the following goal in our hypothetical plan related to environmental risk.

To minimize the risks to lives and property due to landslides and other nonseismically induced geological hazards within the planning area.

This goal may be satisfied in a number of ways. One policy introduced in the plan recommends the following.

Policy 2: Prohibit the construction of any structure intended for human occupancy in any landslide-prone area unless geologic investigations or project mitigation demonstrate low levels of acceptable risk at the site.

Several different programs, including the following three may implement this policy.

Program 2.1: Require geologic and geotechnical engineering studies for all new development prior to the issuance of building permits on slopes greater than 20% and within areas of high, moderate-to-high, or moderate potential for landsliding.

Program 2.2: Require developers to include drainage, erosion, and landslide mitigation measures where necessary to reduce landslide potential.

Program 2.3: Minimize earth-moving activity in areas of moderate to high landslide potential.

The programs identified here direct attention to the use of site investigation measures, mitigation techniques, and construction practices, along with detailed environmental data to reduce the risks associated with landslides. Using these tools, future development will be placed where risks are acceptable and damage to property can be kept at a minimum.

Finally, let's examine a land-use goal with an environmental focus.

To balance housing development and environmental protection.

This is an extremely general goal without much in the way of specifics to help frame policy. However, the idea of balancing development with environmental protection is common. To reach this goal planners have recommended the following policies.

Policy 3: Preserve and enhance environmental quality in conjunction with the development of housing.

This policy statement is tied directly to the following program.

Program 3.1: Require environmental review of development proposals to determine the significance of their probable effects.

Policy 4: Encourage energy and water conservation designs and features in residential development.

Program 4.1: Consider building orientation, street layout, lot design, landscaping, and street tree configuration in subdivision review for the purposes of solar access and energy conservation.

Each of the goal and policy statements presented in the examples identifies a problem, fact, or issue that affects the planning area. Once identified, one or more directives decision-makers can follow to address each problem are presented. Finally, one or more programs are recommended for implementation which over time will satisfy the goal. Above all, these hypothetical goal statements serve to further illustrate the point that all plans share three essential ingredients: (1) facts, (2) goals, and (3) recommendations. (See Fig. 3.2.)

It is important to recognize, however, that plans are never permanent, and they are not the single purpose of planning. As Kaiser et al. (1995) remind us, plans must be updated periodically to reflect changes in conditions and community values, and they must always be related to other community actions. The goals expressed in a plan emphasize a vision of the future and a means to attain that vision. Through the interrelated activities of search, analysis, synthesis, and selection, the data and opinions that drive the planning prob-

Facts
Goals
Recommendations

Fig. 3.2 Underlying plan concepts.

lem can be blended together and formulated into a coordinated statement that details community aspirations and concerns. Articulating these aspirations in the form of a plan demands more that the technical expertise of the planner, it requires the continuous involvement of the public.

The role of the community

Citizen participation in planning remains a widely discussed and debated topic (Day, 1997). In general, citizen participation in the planning process is seen as a positive feature since it provides an important avenue for the planner to elicit community attitudes and values. It also facilitates the creation of a forum for citizens to voice specific concerns and problems that can become the focal point for the development of planning goals. At the same time as citizen participation seems to hold a sacrosanct role in democratic political culture, the issue of public participation in the planning process seems problematic (Day, 1997). This is partly due to countervailing forces in political culture that doubt the ability of the general public to constructively contribute to governance (Stivers, 1990).

These contrasting positions stem from a perceived tension between two groups: those who view planning as a rationally organized activity that places importance on technical expertise and impartiality; and those advocates of democratic social and political systems that contribute "noise" and contradicting beliefs, needs, and perceptions which otherwise confuse this rational process. Consequently, citizen participation and its importance in planning tends to fade in and out of favor. This suggests that in some cases meaningful participation might be conceived as problematic (George, 1994; Day, 1997).

Reconciling these opposing views of the public-participation issue depends on one's definition of planning and how the public fits into this process. A more considered view of planning recognizes that good plans spring from the community. From this perspective, the planner serves to facilitate the planning process and lend expertise. Several points can be made in favor of this ideation (Kelly & Becker, 2000; Levy, 1997). First, it avoids the elitist view of the planner as technical professional who remains aloof and detached from the problem. While the planner has skills that a typical member of the public does not possess, it should not imply that the planner is necessarily wiser (Levy, 1997). Secondly, planning problems are complex and multifaceted. It is therefore unlikely that the planner or any other individual or group can have a complete or accurate understanding of the needs or issues confronting the community. Only by taking the public into consideration and tying them into the planning process can their interests be fully represented. Here, as Levy (1997) notes, a plan formed with community input is more likely to be carried out than a plan of equal quality that has been created only by professionals. Thus, by taking the public into the planning process at an early stage, issues critical to the public can be represented in the plan, and people will be better informed regarding its important details. For example, in Athens, Ohio, efforts have been underway to engage the public as the community reshapes its vision. Public meetings and forums that allow community groups and individuals to voice their ideas and concerns will ensure that the vision formed through planning reflects the shared ideas of the community. Also, with community involvement, a level of commitment to the plan is produced which only enhances the plan's long-term viability. Therefore, there are several reasons why citizens should have the opportunity to participate in planning. The most important is simply that our system of government gives citizens the right to have a strong voice in all matters of public policy, including planning. A second reason is that only citizens can provide the information needed to develop, main-

tain, and carry out an effective comprehensive plan. Professional planners and local officials need comments and ideas from those who know the community best: the people who live and work there. Third, citizen involvement educates the public about planning and land use. It creates an informed community, which in turn leads to better planning. Fourth, it gives members of the community a sense of ownership of the plan. It fosters cooperation among citizens and between them and their government. That leads to fewer conflicts and less litigation. Finally, citizen involvement is an important means of enforcing our environmental and land-use laws. Having citizens informed about planning laws and giving them access to the planning process ensures that the laws are applied properly.

Fostering citizen participation begins first by trying to define precisely who represents the public. One major issue when encouraging public participation stems from the observation that the outcomes of participatory processes do not always reflect the aggregate of citizen preferences or interests (Day, 1997). An all-too-familiar reality is the fact that too few people take advantage of their opportunities to participate. As a consequence, outreach on the part of the planner is a necessary step in the formulation of plans and programs if the public is to participate in their development. A variety of approaches can be offered to encourage wider public involvement. These include the use of:

- advisory panels and committees
- open meetings and forums
- press releases and media coverage
- public surveys and questionnaires
- citizen and neighborhood groups
- public presentations and speaking engagements.

Some combination of these approaches will facilitate dialogue between the planner and the community. However, no method is perfect, and many groups may perceive their needs to be under-served.

Perhaps the best way to have strong citizen involvement in planning is to have strong planning for citizen involvement. In other words, a successful citizen involvement program must be carefully designed and managed. Establish objectives. Assign responsibilities. Allocate money and staff. Set a schedule. Monitor performance. These are basic steps to successful management of any program. Yet all too often these steps are forgotten in the case of citizen involvement. For some reason, citizen involvement frequently is not seen as a program to be actively managed. Rather, it is treated as a passive process, one that will somehow happen automatically if a few notices are mailed and a hearing is held.

It should be recognized, however, that citizen involvement doesn't just happen. The most widespread public participation in planning is found in those communities where involvement is planned and managed carefully and aggressively. Some techniques communities are using include:

- Managing citizen involvement in the same way as code administration or long-range planning – that is, as a major element of the planning program.
- Drawing up a citizen involvement plan for each major legislative action and for land use decisions that involve important community issues.
- Developing a Committee for Citizen Involvement that can:

  advise planners and policy-makers on how to manage citizen involvement for specific projects,

  periodically evaluate the citizen involvement program,

  work with staff to maintain an effective network of citizen advisory committees,

  act as a mediator to resolve disputes about public participation,

  act as an ombudsman for citizens concerned about public participation.

- Staffing the citizen involvement program with a professional coordinator from outside the planning department. This arrangement has several advantages. It frees planning staff from citizen involvement duties that might conflict with or take second place to other planning tasks, such as code enforcement. It allows for broader community involvement: citizen concerns are not limited to land use. And the coordinator can serve as a mediator if the planning department and citizen advisory committees disagree about a land-use issue.

- Giving planners who deal with the public training in customer relations and communications.
- Using role-playing and simulation exercises to help planners, planning commissioners, and other officials to understand the needs and wants of citizens and interest groups.
- Maintaining a registry of stakeholders, interest groups, and individuals with expertise or interests in important land-use topics or areas. Use that registry as a source of contacts when deciding whom to involve in a particular citizen involvement effort. Update the list periodically.
- Earmarking funding for citizen involvement in the budget. Goal 1 requires this, and for good reason: it helps make people aware that citizen involvement cannot happen without a commitment of resources.
- Developing and maintaining an active network of neighborhood organizations. Make sure the committees continue to receive information about permit applications, policy issues, and major projects, such as revisions to the plan or development codes.
- When seeking members for a key committee, using an open process, such as published notices, contacting local civic groups, and posting announcements.

When one is confronting environmental problems, citizen participation in political, community, and neighborhood affairs is critical to the creation and maintenance of a strong, vibrant community. A community without regular interaction among citizens is less a community than a random collection of people. Without active participation, it is difficult for a community to agree on what problems to address and how to move forward collectively to solve them. This means that citizens must be engaged in decision-making processes from the beginning (Kelly & Becker, 2000; Hanna, 1995; Barber, 1981). Two useful instruments to foster early involvement include neighborhood meetings (Table 3.6) and public hearings (Table 3.7). To encourage participation,

Table 3.6 Basic guidelines for making a presentation at a public hearing.

1 Keep in touch with the planner assigned to the item. The planner can notify you of postponement or new information.
2 Speak to the issue at hand. If a zone change in a master-planned area is being considered, address the merits of the request and not, for example, whether the master plan should have been approved initially.
3 Give letters, petitions, and other documentation to the assigned planner *before* the hearing. The planner will distribute the material to the members of the public body.
4 While there is usually no time limit on comments, be brief and to the point; do not repeat comments made by others.
5 If many people are interested or intend to speak on the item, you may want to select one or more representatives to give the group's position. Anyone wishing to address the item, however, may speak.

Table 3.7 Basic guidelines for participation by neighborhood groups.

1 Keep in touch with the planner in charge of your geographic area.
2 Provide City Planning with a current contact person and phone number.
3 Invite planners to come to your meetings to discuss issues of specific interest.
4 Prepare a map showing the boundaries of your neighborhood group. If it is provided to City Planning, it will assist efforts to notify the group of proposals in the area.
5 Take the time to understand the ordinances and the process.

it's essential that citizen groups: keep in touch with the planner in charge of their geographic area, provide the planning staff with a current contact person and phone number, invite planners to come to meetings to discuss issues of specific interest, prepare a map showing the boundaries of the neighborhood group, and take the time to understand the ordinances and the process.

Of particular concern when working with the public participation question are those problems or issues that are highly controversial, and poorly defined. In this case, identifying goals, understanding the problem, and outlining any course of action can be made difficult by the complexities and uncertainties surrounding the problem. The situation can be made worse by public perception

and the diverse attitudes and opinions that must be reconciled. To illustrate this point, consider the dynamics surrounding a locally unwanted land use. The terms NIMBY (Not In My Back Yard) and NOPE (Not On Planet Earth) are very familiar positions when dealing with controversial, high-risk, or poorly understood environmental problems. Both define very strong perceptions that can frustrate the planning process. While they can prove to be intractable positions, they have to be addressed. If not, they only serve to intensify public mistrust of the planning process and encourage an irreverence for the official version of reality that may be offered by the planner (Fischer, 1993).

Planners who view their role as technocratic can inadvertently encourage contrasts in perception; one can be conferred a special status by her or his peers after demonstrating a mastery over a technique or body of knowledge. The attitude which accompanies this ability to calculate unequivocally correct and precise answers, excludes the public from participation and imposes barriers based on scientific knowledge and technical jargon (Van Valey & Petersen, 1987). When opportunities for public participation are provided, the primary mechanism takes the form of a public hearing. However, such hearings often confound and discourage participation (Day, 1997). Public hearings are criticized for:

- occurring too late in the decision-making process.
- being scheduled at times that are inconvenient for the public.
- establishing an atmosphere that inhibits dialogue.
- conducting proceedings that intimidate the public.

The perceptions created by a technocratic philosophy encourage NIMBY activism and further polarize community interests. The resulting deadlock frustrates the planners' ability to achieve consensus on critical issues and contributes to antiparticipation attitudes (Morris, 1994; Inhaber, 1998). Therefore, rather than defining NIMBY attitudes as an irrational response to problems ordinary citizens cannot grasp, the simple solution is to encourage more citizen participation, not less (Fischer, 1993). In fact, Fischer maintains that planning analysis should be viewed as an evaluation of alternative solutions employing criteria derived consensually. This view suggests a collaborative approach to plan formulation that emphasizes direct contact between those promoting locally unwanted activities and all affected parties (Dear, 1992).

Accepting the premise that the planner's primary obligation is to serve the public interest, the challenge for planning is how to effectively integrate technocratic and democratic contributions when addressing complex issues. Five guiding principles have been recommended (DeSario & Langton, 1987):

1 that the dangers associated with maximizing expert and citizen contributions without joint review and interpretation be avoided.

2 that the unique contributions of experts at the technical level and of citizens at the normative level of policy-making be encouraged, but that a later stage of mixed review be created that involves experts and citizens in examining issues of impact and trade-offs regarding technocratic and democratic considerations.

3 that the issue of the role and power of citizens be made explicit at the outset, and appropriate procedures be developed to reflect power-sharing arrangements.

4 that adequate information, access to it, and technical resources be made available to citizens.

5 that government be experimental in selecting, evaluating, and refining the procedures for integrating expert and citizen contributions that are most effective in dealing with the unique policy issues with which each are concerned.

With citizens fully engaged in the process, focus can be directed at developing the environmental plan.

Developing environmental plans

Formulating plans and developing a format that compartmentalizes the planning area into specific themes provides a structure that facilitates inclusion of all the relevant information needed for effective policy-making. This technical aspect of planning, focusing on the guidelines to follow when preparing plans and the procedures that carry the process through to completion, have been discussed in excellent detail by Anderson (1995). Technical guidelines typically direct attention to the comprehensive plan. Although this is a reasonable place to begin, the environmental planning problem is unique. While many aspects of the environment are discussed in a comprehensive plan, there are features of the environment that warrant special consideration. These are the goals and objectives specific to the environment that can be sufficiently different from those surrounding land-use and development issues that they need to be treated as such.

The concept of an environmental plan or comprehensive environmental plan is not a new idea (Miller & De Roo, 1997). A comprehensive environmental plan is a mechanism communities can use to meet the present responsibilities of environmental protection and the future challenges of enhancing environmental quality. With is exclusive focus on the environment, the environmental plan becomes an important way for a community to:

1 Set environmental priorities and establish clear goals and objects targeted toward environmental issues.

2 Identify environmental resources important to sustainable development.

3 Provide a blueprint for compliance with environmental regulations that affect the community.

4 Explore alternatives to prevent pollution and efficiently manage environmental resources.

5 Develop community support and awareness to environmental protection needs.

6 Create an environmental infrastructure that complements community well-being.

The rationale for producing an environmental plan is essentially a response to the more typical tendency to treat the environment in pieces. Thus, while we recognize that environmental variables are interconnected, our general approach to the environment is to fix symptoms one at a time (i.e. protecting water, protecting air, protecting land). With the growing recognition that environmental protection cannot be successfully achieved unless the environment is treated as an integrated whole, the community environmental plan becomes a central instrument for that integration. However, for an environmental plan to be effective, the environmental responsibilities of the community need to be understood. These responsibilities extend well beyond the larger environmental issues of population, resources, and pollution, or the immediate controversies that grow out of landscape development pressures. The environmental responsibilities germane to the environmental plan are those of local concern where local resources can be committed to their improvement:

- drinking-water quality
- wastewater management
- solid waste management
- leaking underground storage tanks
- hazardous waste management
- emergency response to hazard
- groundwater protection
- wetlands protection
- flood plain zoning
- risk assessment and management
- pollution control.

With attention given to the local environment, developing comprehensive environmental plans shares many similarities with the general procedures outlined previously. The main difference separating environmental plans from land-use or general plans is one of focus. Environmental plans are directed toward the environmental challenges that face the community. To meet these challenges emphasis is given to five main phases in crafting the plan (see Fig. 3.3):

1 Developing an environmental vision.
2 Defining environmental needs.
3 Identifying feasible solutions.
4 Implementing the environmental plan.

We will explore these plan formulation stages in the sections that follow.

Developing an environmental vision

With attention directed at the local community, developing an environmental vision serves as a framework that assists all parties involved in making choices about environmental goals. This

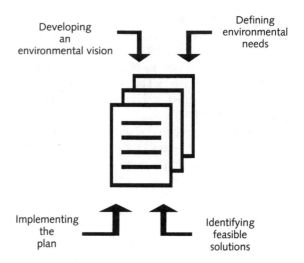

Fig. 3.3 Features of the plan formulation process.

Table 3.8 Questions that characterize place.

Who makes up the population?
What is unique and important about the planning area (socially, culturally, historically)?
What are the strengths and weaknesses of the local economy?
What are the important characteristics of the natural environment?

vision takes shape in response to a series of questions that focus the goal-setting process. Examples of questions that may be posed during this phase include:

- *What features characterize the planning area?* This simple question addresses several issues related to establishing an understanding of place. Examples are given in Table 3.8.
- *What are the community's attitude and values relative to the environment, economic growth, and lifestyle?* Communities have different attitudes toward development: some prize stability and traditional ways of life, while others consider growth and change to be important to community survival. In either case, planners need to understand the extent to which qualities such as environmental preservation, growth, and development are valued by the community.

Table 3.9 Targets of change.

Target	Screening questions
Natural environment	Are there trends in the loss of natural resources that should be reversed? What can be done to protect resources and prevent pollution?
Land use	Is the current mix of land used a good balance? Should some areas be used differently? What trends define the region?
Infrastructure	What level of service should be provided? What is the age and capacity level? Will infrastructure accommodate growth?
Demographics	What are the important demographic trends? What level of growth can be supported?
Economic growth	How will growth influence the quality of life? Is there a need to attract growth? What type of growth is desired?
Community concerns	Are there health or risk issues that need to be addressed? How does public health compare to other regions?

Table 3.10 Approaches used to define boundaries.

Town or village boundaries
Service area boundaries
Special districts
County boundaries
Physical characteristics

- *What changes or improvements within the community are desirable?* Here, a range of possible areas of change may be examined and opinion may be elicited for each. A selection of those pertinent to environmental planning is provided in Table 3.9.

Defining community needs

The answers obtained from the questions relating to the community's environmental vision begin to focus on specific needs. Need, in this context, explains those features of the environment that are of greatest concern. Crystallizing an understanding of need and expressing need as a geographic feature of the planning area requires the planner to:

- Establish the boundaries of the environmental planning area.

- Review the environmental regulations that affect the planning area.
- Identify the environmental problems that threaten environmental quality.
- Evaluate the effectiveness of existing environmental management facilities and infrastructure.

Bounding the planning area Delineating the planning area concentrates efforts on including problem areas that have actual or potential public-health and ecological impacts, critical resource areas that serve the community and require protection and preservation, and facilities used to protect public-health or environmental qualities. Boundaries can be defined in several ways. Some of the more common methods are outlined in Table 3.10.

Existing environmental regulations There are numerous state and federal regulations aimed at an array of environmental factors. These regulations can influence and help define key environmental needs. In fact, not only will these regulations help identify environmental issues that must be incorporated into the plan, but they also define standards and compliance measures that if not addressed may result in fines and penalties. Table 3.11 describes regulations that can be used to address best management practices in the plan.

Identifying existing environmental problems This step in defining need concerns efforts to devise a listing of any and all environmental problems that represent serious threats to health and ecosystems in the planning area. Possible threats may include unsafe drinking water, specific pollutants or pollution sources, or natural resources that may be affected by pollution. An sample list is shown in Table 3.12. Critical to compiling this list is the

Table 3.11 US environmental regulations to address management practice.

Drinking water quality	Safe Drinking Water Act
Wastewater treatment	The Clean Water Act of 1977
Wetlands protection	Clean Water Act – Section 404 Food Security Act – Swampbuster Section
Nonpoint source pollution	National Nonpoint Source Program Clean Water Act
Solid waste management	Resource Conservation and Recovery Act
Hazardous waste management	Resource Conservation and Recovery Act
Emergency response to hazardous substances	Emergency Planning and Community Right to Know Act (SARA Title III)
Asbestos-containing materials	Asbestos Hazard Emergency Response Act
Radon gas	Indoor Radon Abatement Act
Air pollution	Clean Air Act

Table 3.12 Environmental problems checklist.

Air quality
Asbestos in buildings
Chemical hazards/releases/spills
Drinking-water quality
Ecosystem/habitat quality
Flood hazard
Hazardous waste sites
Nonpoint pollution
Pesticides
Radon
Solid waste
Surface water
Underground storage tanks
Wellhead/watershed protection
Wetlands

planner's ability to assess the level of risk each item may represent in relation to environmental quality. A series of screening questions can be used to help define risk and highlight problems that may require more detailed investigation. Examples may include:

- What harmful effects can the substance or activity cause to human health or ecosystem functioning?
- Are these effects permanent or reversible?

- What are the effects of different levels of exposure?
- To what degree is the local population or ecosystem exposed to the substance or activity in question?
- Is there presently any evidence of harm to human health or ecosystem functioning as a consequence of exposure?
- What are the known concentrations of the substance in critical receptors?

Effectiveness of existing facilities An important step in identifying need involves the critical evaluation of the community's environmental facilities. Here, consideration is given to facilities such as landfills, incinerators, transfer stations, recycling centers, water treatment plants, wells, wastewater collection and treatment centers, as well as buffer zones, wet ponds, and swales for runoff management. The purpose of this review is to determine whether a facility is performing effectively and is capable of meeting present and future demand. Inadequate performance may indicate that operations may be functioning beyond carrying capacity and require modification. Therefore, evaluating facility performance helps to (1) identify potential risks and (2) determine whether the community is in compliance with local, state, or federal regulations. A series of screening questions to help review facility adequacy is given in Table 3.13.

Following the careful definition of community need and after the environmental vision of the community has been refined, attention can shift to the consideration of possible solutions and strategies for enhancing environmental quality. The list of possible solutions and their integration into the comprehensive environmental plan is examined in the sections to follow.

Identifying environmental solutions

There are a variety of options for achieving environmental goals. To determine which if any are suitable to the given problem, information is needed in order to ascertain

- What each solution can achieve.
- What factors limit a solution's effectiveness.
- What the costs associated with the solution are.

Table 3.13 Facility effectiveness screening questions.

Design concerns

Is the plant design adequate for the existing demand?
Will it accommodate future demand?
Does the facility meet requirements of current
  regulations?
Are maintenance problems increasing?

Management concerns

Is management clear about the system's goals?
Have managers evaluated present and future levels of
  service?
Is the facility adequately staffed?
Does the staff understand their responsibilities?
Are the revenues generated sufficient to meet current
  service demands?
Are funds being set aside for improvements and
  expansions?

Operational concerns

Are facilities operating at or near capacity?
Can the facility adjust to changes in input type and
  quantity?
Are mechanisms that control processes in good operation?
Have potential hazards been identified?
Have operating procedures been updated?
Are good records kept?

Table 3.14 Planning solutions to common environmental problems.

Environmental concern	Solution alternative
Drinking-water quality	Protecting the source Improving treatment technologies Point of use/point of entry fixes
Drinking-water quantity	Conservation Leak detection Identifying new supply sources
Wastewater treatment	Use of onsite systems Cluster systems Centralized systems
Solid waste	Source reduction, recycling, composting
Hazardous waste	Household hazardous waste collection programs
Nonpoint pollution	Identifying sources Developing management strategies Educating the community

- How the solution can be implemented.
- Whether the solution will affect of contribute to other environmental problems.
- Whether the solution will foreclose on future options.

With preliminary answers to these questions, a list of generic options for key environmental issues confronting the planning area can be created. Options can be eliminated based on their degree of feasibility, but no reasonable option is omitted from consideration. For example, options may be eliminated that will not work because of situations specific to the planning area. These may include:

1 population/demographic characteristics
2 distance constraints
3 local hydrography/topography
4 soil and geologic conditions
5 environmental chemistry.

Options may also be eliminated due to cost considerations. Cost may be defined simply as a matter of economic factors, although other expressions of cost should be examined (i.e. social costs, ecological costs, health-related costs). Finally, options may be eliminated because they require

advanced technical skills beyond that which can be accessed in the community, or because they are too complicated to be administered successfully.

After a list of possible options has been reduced to a more feasible set, the planner must review each and clarify precisely what each solution will achieve. During this review and evaluation, several of the remaining solutions may be rejected. The strategy to remain focused on is to never remove a potential solution without an assessment of its performance capabilities. In many situations it may be necessary to employ a combination of solutions, and frequently different solutions can complement one another and net an overall greater benefit. A list of generic solutions arranged by environmental issue is present in Table 3.14.

Prioritizing objectives

One of the more critical steps in developing a comprehensive environmental plan involves the task of targeting the most important problems that it should address. While this sounds comparatively straightforward, setting priorities requires a willingness to trade-off development objectives such as attracting businesses into the community or promoting tourism for those related to environmental quality. A simple means of establishing a ranking of objectives places each problem into a subjective categorization based on relative risk:

1 **Urgent** – identifying those environmental problems that present the highest risk to human health.

2 **Necessary** – defining problems with lower levels of risk and where regulatory violations exist.

3 **Desirable** – describing problems that present no regulatory concern and exhibit low levels of long-term risk.

Once the major objectives have been selected and priorities established, the plan can be formalized into a detailed statement of the community's environmental vision. The planning process can now shift focus to consider the issue of implementation. Implementation is the mechanism whereby the plan can be put into action, its performance evaluated, and its focus revised as conditions and needs change.

Plan implementation

The most carefully crafted plan may never achieve its designed effect simply because it lacked a strategy for implementation. For example, a plan for a highway bypass to reduce traffic congestion through a town generally requires selecting a route, purchasing right of way, conducting environmental reviews, designing the highway, planning for construction, and acquiring funding. To implement this plan requires a strategy to ensure that everything takes place in the sequence necessary to produce success. In this sense implementation explains a set of procedures that can be used to put our environmental plan into practice. A general outline detailing one possible strategy to guide implementation has been offered by Anderson (1995). This procedure consists of ten steps (see Fig. 3.4):

1 Review the goals, policies, and recommended actions in the plan to identify:

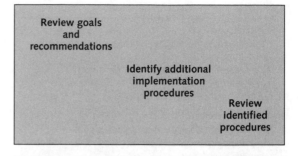
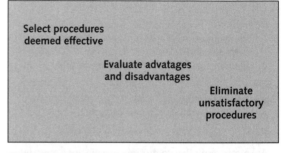
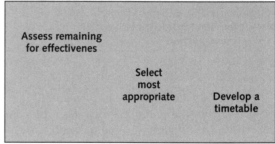

Fig. 3.4 Guidelines for plan implementation.

- those with satisfactory procedures already in place that can guide implementation.
- those which cannot be implemented presently due to their general nature.
- those which should not be implemented at present due to legislative, economic, political, or technical constraints.
- those which will require new or revised implementation procedures.

2 Identify possible additional plan-implementing procedures that can be reasonably entertained.

3 Conduct a preliminary review of the procedures identified.

4 Select procedures that will produce desired and effective results.

5 Evaluate the potential procedures, soliciting comments on their relative advantages and disadvantages.

6 Eliminate procedures that are considered ineffective, politically unacceptable, too complex, or too costly to institute and administer.

7 Prepare an analysis of the remaining procedures to ascertain:
   • what the objectives of the program are.
   • how effective the program will be.
   • what the administrative requirements are.
   • who would be adversely affected by the program.
   • what the probable benefits of the program are.
   • what legal steps are required to enact and administer the program.

8 Review the findings from Step 7.

9 Select those procedures that appear to be the most appropriate.

10 Develop a timetable for the introduction, adoption, and administration of the selected procedures.

Because long-range environmental plans are general and in some instances purposely non-specific, not all plan implementation programs are suitable for use without some modification. A generalized set of instructions for implementing long-range comprehensive environmental plans would give emphasis to (after Anderson, 1995):

   • Selecting and using those plan-implementing measures that are clearly suitable to chart a course of change over a 20-year period.
   • Preparing short-range programs that are specific enough to guide change using 5-year increments.
   • Implementing short-range programs using procedures suitable for the immediate future.

Implementation methods and measures

There are two broad categories of action that can be taken to implement programs and policies expressed in a plan (Levy, 1997): (1) public capital

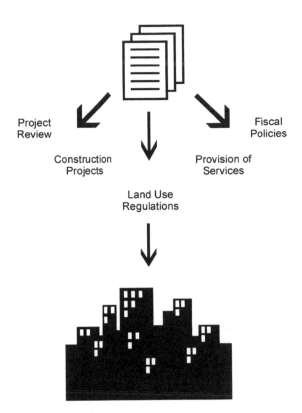

Fig. 3.5 Tools for plan implementation.

investment, or (2) public-control land utilization. From either of these two directions, several types of plan implementation programs can be developed (Fig. 3.5). Examples may include

   Construction of physical facilities
   Provision of services
   Regulation of land use and development
   Project review
   Fiscal policies.

Programs, such as zoning ordinances and subdivision regulations, are the traditional tools of the planner; however, a wider mix of programs is more typical. Examples of these broader strategies can include public land acquisition programs, housing and development programs, redevelopment, capital improvement programs, and the use of transferable development rights. For instance, when the plan to implement the Mid-Peninsula Regional Open Space District was undertaken in California, a public land acquisition program was one strategy used to realize the goal of providing access to open space. Several of the more com-

Table 3.15 Land-use factors subject to zoning.

Type of land use
Activities permitted on private properties per land-use type
Minimum lot size
The physical placement and spacing of structures
Maximum percent of lot covered by structure
Maximum building height
Amount and design of offstreet parking
Design of structures and sites
Minimum/maximum floor area
Permitted noise levels
Design review

monly used implementation tools are described below. A more detailed treatment of these methods can be found in Kelly and Becker (2000), Levy (1997), and Anderson (1995).

1 Public regulatory/land-use controls   Zoning defines the delineation of the planning area into districts and the establishment of regulations within these districts to control the type, density, spacing, and placement of permitted land uses. Zoning ordinances typically include provisions that regulate site layout, structural characteristics or buildings, and procedural actions pertaining to compliance and zoning appeals. A general listing of the land-use factors subject to zoning control is provided in Table 3.15. As a form of control, zoning is an attempt to avoid disruptive land-use patterns and prevent the location of activities within districts that may generate external effects that may be detrimental to existing or future land uses. The concept of zoning can be extended to include the regulation of uses that may impose significant environmental risks or describe conditions that are environmentally incompatible.

Subdivision regulations define any ordinance adopted or administered by local government which regulates the division of land into two or more lots, tracts, or parcels for the purpose of sale, lease, or development. Subdivision regulations give communities power to ensure that new residential development meets community standards and complements the goals and objectives of the comprehensive plan. A subdivision ordinance will specify the administrative procedures to be followed in the division of land, design standards that must be adhered to, and identify the improvements that must be installed such as streets and utilities. The ordinance will also regulate the manner by which parcels of land may be converted into building lots, and stipulate which improvements must be made before building lots can be sold or building permits granted. Ideally, subdivision regulations are designed to meet several purposes, and while most are targeted toward residential development, many of the same concepts and controls can be used to govern commercial and industrial subdivisions as well.

Growth management programs describe programs prepared, adopted and administered by local government that are designed to regulate: (1) the amount of urban growth, (2) the rate of urban growth, (3) the type of growth, and (4) the location and quality of growth. Perhaps the most well known growth management program was that used by the city of Petaluma, California, nearly three decades ago. Such programs are intended to discourage or severely constrain unwanted urban development, particularly in situations where growth would:

1 Change the characteristics of the community.
2 Be detrimental to the economic base of the planning area.
3 Generate loads that would strain or exceed carrying capacities.
4 Produce adverse secondary environmental impacts.

Design review describes procedures developed to facilitate the review of proposed building designs and regulate the site and structural characteristics. In most cases, review focuses on the physical design of individual structures, historic districts, office parks, and industrial sites for individual buildings or groups of buildings.

Impact assessment reports explain a set of procedures followed to analyze and disclose the potential effects of a proposed action or project on the local environmental system. Impact reports may concern purely environmental consequences or they may be broadened to include social, economic, and fiscal impacts of the action as well. In each case the purpose of assessment is to identify the potential short-term impacts of a proposed project so that an informed decision concerning the adverse and beneficial effects of an action can be made. The assessment procedure followed is intended to: (1) identify all relevant adverse and beneficial effects, (2) identify mitigation measures that would reduce adverse effects, and (3) identify alternatives to the proposed action (Canter, 1996).

2 Public capital investment programs It may be argued that accessibility is the main determinant of the development potential and value of land (Levy, 1997). The nature of accessibility to the landscape and the value of land are heavily influenced and shaped by public investment in critical infrastructure. Public investment in infrastructure may take many forms, but most typically the term applies to government provision of roads, parking facilities, sewer and water lines, and related facilities. Indeed, one of the more useful means a community has to direct its growth is by investment (or unwillingness to invest) in road construction, sewer lines, and water hook-ups. Public investment programs can include a variety of instruments used by government to extend or withhold infrastructure improvements within the planning area. Examples include the following.

Public construction projects are sponsored by a public agency and designed to create or improve roads, transit systems, public buildings, or water and sewage systems.

Public land acquisition programs explain the purchase of land in fee simple or the purchase of limited rights to land in order to: (1) make a site available for full public access and use, or (2) acquire limited rights to property such as water rights for lands draining into a reservoir, development rights for open space districts, or air rights for land adjacent to airports.

Economic development programs define activities intended to generate wealth by mobilizing human, physical, natural, or other capital resources to produce marketable goods and services. Such programs attempt to foster economic growth, provide employment opportunities, and develop a strong tax base through the creation of mechanisms to (1) retain existing businesses and industries, (2) attract businesses, (3) nurture small businesses, and (4) develop facilities that capture businesses.

Housing programs are created to provide housing for residents within the planning area by implementing policies, strategies, and proposals that encourage (1) occupancy by the type of occupant, (2) occupancy by the income of occupant, (3) need-appropriate housing types, (4) home ownership, and (5) alternate patterns of location.

Transferable development rights define the transfer by sale or barter of some or all of the right to develop a parcel of land located in one district to a parcel of land located somewhere else in the planning area. The concept of transferring rights to development has been used to allow the sale of (1) "air rights" over historic areas and buildings as a preservation measure and, (2) development rights in rural areas where development is unwanted.

Plan evaluation

Do plans work? This and a series of interesting questions have been raised by Talen (1996) on the topic of plan evaluation. Since the success of plan-making can be determined only at some future point in time, the question as to how planners evaluate whether or not the plans they create are actually implemented or whether their plans ever achieve their desired objectives is anything but trivial. Evaluation in planning, however, is complex, and embodies a variety of instruments and methodologies (Talen, 1996). To help distinguish between the evaluation of plans and other types of evaluation undertaken in planning, we can separate the concept into four main categories (Talen, 1996):

1 Evaluation prior to plan implementation.
2 Evaluation of planning practice.
3 Policy implementation analysis.
4 Evelution of the implementation of plans.

According to this outline, category 4 defines those procedures and instruments that focus on the question of implementation. Here, evaluating how well the plan is working can follow either of two paths:

a) Qualititative approaches – these methods and instruments employ evaluative mechanisms that are subjective and selective in nature. An example might be the annual review of building permits to see how well the permit process agrees with provisions in the plan to encourage more multifamily housing. By looking at the pattern a judgment can be made as to whether the community is reaching this goal. Using this approach, evaluative criteria can be compared against objective indicators of success, such as levels of economic well-being or related ideas.

b) Quantitative approaches – are methods that rely on empirical investigations and quantitative support to determine the success of a plan. Examples of this approach may include the systematic sampling of key water quality indicators downstream from the municipal sewage treatment facility to see if investments made in new water-treatment technology have led to improved water quality. Two general approaches have been found to be useful in this regard: (1) the use of map overlays to quantify the level of agreement between the actual form of development and what may have been suggested in the plan, and (2) the use of inventories to document measurable relationships between attributes used to characterize aspects of the planning area (Alterman & Hill, 1978; Calkins, 1979). Additional approaches to quantitative plan evaluation have been presented by Bryson (1990), and Kartez and Lindell (1987). Most of these methods use regression-based approaches to establish the degree of fit or correlation between actual outcomes and those specified in the plan. In the method described by Bryson (1990), success of outcome is subjectively scaled, as are the explanatory variables used to evaluate implementation. Using these scores, the results of the regression model present the relationship between goal achievement and (1) successful problem identification, (2) conflict resolution, and (3) impact on resource allocation.

The methods reviewed above suggest that for evaluation to be successful, evaluative criteria must be carefully selected and defined. Useful evaluative criteria are those attributes of the planning area or problems that can support the detection and measurement of change, provide a means to define success, facilitate the analysis and treat-

ment of issues of multicausality, and illuminate expected outputs. Above all, since plans are formulated with the intent of being implemented, an evaluative component must be part of the planning process to provide feedback as to how well the process is working. The key to integrating a dynamic evaluation component into the plan rests on the planner's ability to (1) incorporate evaluative methods explicitly, and (2) provide a means to measure the achievement of each goal as an integral part of the plan. As Talen (1996) maintains, once planners know what elements of plans are successfully implemented and what elements are not, they can move quickly to the next tier of evaluation. This aspect of evaluation directs efforts toward the identification of the underlying factors associated with successful plan implementation and where things went wrong. Systematic failure to meet the goals expressed in a plan may indicate that the community is pursuing the wrong goals (Talen, 1996).

Summary

Planning becomes embodied in a plan. The nature of that plan was described in this chapter. Specifically, the plan, defining the goals, objectives, and policy recommendations of the planning area, was examined as both a physical document with clearly identified elements that frame community aspirations, and as a program for the future that describes a future state of the region and how its arrangement will take form. In either regard, plans contain information. The type of information required and how this information is presented was reviewed. However, a plan will never achieve its goals unless it can be implemented. In this chapter the basic mechanisms available to implement plans were examined and the larger question of evaluating the success of a plan was discussed.

Focusing questions

How might the time horizon established for a plan influence its success?

Describe the intellectual tools needed to draft and produce a working plan.

Discuss the role of citizen participation in the plan-making process.

Explain three important functions a plan is designed to fulfill.

Discuss the use of projection and forecasting tools in plan design and analysis.

CHAPTER 4

Natural Factors in Environmental Planning

Environmental planning has been explained as a systematic attempt to integrate environmental and earth science information into the land-use and land development process. With this information in hand, the environmental planner strives to reduce the adverse effects of human-induced landscape change on both the social and environmental systems characterizing the planning area. The principal goal of this approach to planning is to identify alternatives that maximize human and environmental benefits while minimizing the entropic influences of land development programs. The distinguishing characteristic of the environmental planning approach is its focus on the utilization of environmental information to guide decision-making concerning land development and regional growth and change. In this chapter we will examine the type of environmental information critical to the environmental planning approach, and review the essential natural factors that guide environmental planning and decision-making.

The relevance of environmental information

Environmental information speaks to the myriad factors that define the physical landscape and impart an influence on the planning process. Integrating environmental information into this process, however, requires an understanding and sensitivity to the physical processes and patterns that shape the environmental system. Therefore, assessing natural systems and defining their role in critical environmental processes is a central activity for the environmental planner. Assessment begins with a search for information. Specifically, when natural factors are examined, the planner seeks and uses this information to help answer fundamental question regarding how the environment influences the resource potential of land and the appropriate direction change can take without adversely affecting the landscape. With respect to information search and assessment, for each variable that has been selected to characterize the environmental system, the planner would like to know whether a given environmental factor is relevant to the problem, what it explains, what influence it exerts on environmental and development processes, what specific information is needed to form a complete understanding of its significance, and how it is defined within the landscape system. Answers to these basic questions help to refine the salient features of the landscape and also identify the critical knowledge needed to effectively guide environmental planning. In addition, the natural factors selected for review aid planning by defining important earth-science controls that help to shape a more harmonious balance between human motivations and natural form. As Legget (1973) reminds us, when planning starts, the area to be developed is not the equivalent of a piece of blank paper ready for the free materialization of the ideas of the designer, but rather is an environment that has been exposed for a very long period to the effects of many natural modifying factors. Development of new communities or the charting of regional growth must take into account the fundamental organic and dynamic character of nature so that human progress may fit as harmoniously as possible into this setting.

The role of natural factors in planning

Within the mosaic of the landscape, a natural factor may be thought of as a physical property of the environmental system that connects in some way to a structural attribute of the systems that define human and social processes. This may be as simple as an expansive soil type affecting the integrity of a home's foundation, or may be more complex, such as a regional climate that induces upper-level thermal inversions that degrade ground-level air qualities. Another way to explain these physical and biological factors is to view them as natural resources that evidence the basic ecological premise that "everything is connected to everything else." Regardless of how they are conceptualized, natural factors define controlling variables whose form and consequence direct the balance between land-use activities, facilities, and the environment. Admittedly the environment is difficult to characterize succinctly simply because of its many defining attributes and the complex interrelationships that exist among them. While changes in the attributes that define the environment and their interrelationships describe a new state of the landscape system, there is always an element of uncertainty surrounding the precise implications of this new state and its relationship to basic human needs and expectations, and what change may mean to the future use of the landscape. Therefore, within the interwoven fabric of time, scale, and spatial variability, select characteristics of the environment serve as indicators and regulators of forces set into motion by human actions. From the simple clearing of trees to provide for housing to the accelerated erosion that is washed

Table 4.1 Landscape-shaping forces.

Wave action
Wind action
Glacial Action
Runoff:
- overland flow
- stream flow
- soil moisture
- groundwater

into a river as sediment, the balance of the natural system can be impacted. Therefore, for effective environmental planning, the planner must understand the fundamental characteristics of the landscape and learn how the land functions, changes, and interacts with the life it supports. This point has been eloquently stated by Marsh (1997).

The landscape on which we plan is dynamic. Its form and features are constantly subject to change. As a process, change is driven by system events such as precipitation, soil formation, stream flow, and land use. These actors shape the landscape and direct the planner to not simply respond to patterns and features in a static setting, but to consider the processes and the dynamics of change that they reveal. Through observing the landscape, insight is gained regarding the processes that presently operate there (Marsh, 1997). Changes in these processes alter the functional character of the environmental system, and become recognizable to us as observable changes in landform, climate, soil, hydrology, vegetation. One of the critical tasks of the environmental planner, therefore, is to determine the formative processes that act on the landscape and what these processes mean with respect to present and future use. A listing of these is given in Table 4.1. Because these processes vary both geographically and over time, they produce terrain features that are subjected to differential forces that sculpt, weather, and shape the landform. This is our "blank sheet of paper," a page where the margins have been preset by natural forces that continue to act on landscape.

The patterns which evidence the relationship between natural factors and landscape evolution are held in a dynamic equilibrium as the stress of various driving forces applied to the landscape converges against the inherent resisting forces that strive to maintain balance. For most natural landscape a state of balance exists between the driving forces of water, wind, and human activity and the resisting forces that define the landscape's internal stability (Marsh, 1997). Only when there is an event powerful enough to exceed the strength of the resisting force will the landscape's balance be upset. In the majority of instances, landscapes are resistant to all but the most extreme events. There are exceptions to this general rule, particularly where balance is conditional on a specific feature of the environment. Therefore, recognizing conditional factors present in the landscape is critical to effective environmental planning: these factors induce change. Guiding land development in a manner that will not trigger or compromise the status of these controlling variables is essential. Ignoring them will lead to a series of potentially adverse consequences that can be traced directly back to the changes introduced by land development practices (Marsh, 1997). Therefore, identifying the important natural factors that influence land potential and knowing why they are relevant is an integral part of any plan targeted at achieving sustainable development, and an essential ingredient of the planner's professional knowledge.

A comprehensive discussion of the problem of environmental definition and description can be found in Canter (1996). Although this discussion is directed more toward the issues surrounding the environmental impact assessment process, the basic "what we need to know" knowledge outlined is common to the environmental planning problem as well. In general, the natural factors selected to form a baseline against which planning and change can be evaluated must serve three related functions:

1 They must aptly characterize the environment.
2 They must contain sufficient information to guide planning.
3 They must be pertinent to the goals that motivate environmental planning.

The critical natural factors that must be understood and applied in the environmental planning process may be placed into three general categories with specific attributes listed under each main heading. This outline, or inventory, looks like this:

1 **Physical environment**
   a) Geology
      1) surface and subsurface characteristics
      2) geomorphic controls
   b) Topography
      1) slope form
      2) slope composition
      3) slope stability
   c) Soils
      1) soil type – composition and texture
      2) soil properties
      3) geotechnical characteristics
   d) Hydrology
      1) surface-water characteristics
      2) watershed drainage patterns
      3) groundwater systems
      4) fool plain characteristics
   e) Climate
      1) regional and synoptic patterns
      2) microclimates
      3) extreme events
   f) Hazards
      1) earthquakes
      2) landslides
      3) subsidence
      4) drought
2 **Biological environment**
   a) Terrestrial ecosystems
      1) natural vegetation
      2) natural fauna
      3) community functions
      4) ecological functions
      5) biological resources
      6) sensitive habitats
      7) endangered species
   b) Environmentally sensitive areas
3 **Human environment**
   a) Demographic patterns
   b) Land use
   c) Physical infrastructure
      1) housing
      2) education
      3) public services
      4) transportation
      5) recreation
      6) utility systems
   d) Cultural resources

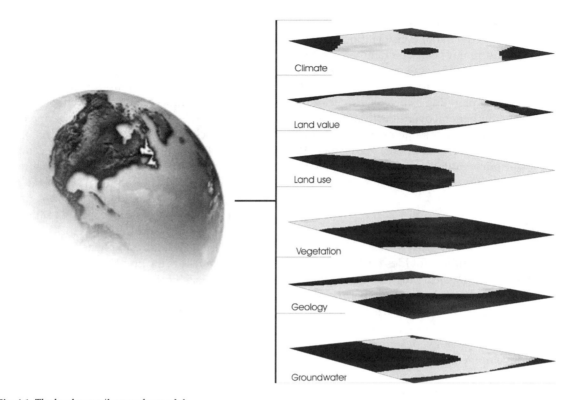

Fig. 4.1 The landscape/layer cake model.

Taken together, the elements listed above define the landscape and can be conceptualized as series of themes or layers that relate and interact over time and space. When assembled into a composite view, these layers, as illustrated in Fig. 4.1, help to define the morphology of the surface and describe its form and functional characteristics. Through assessment these functional qualities become known, and supply the foundation information from which plans are made.

Assessing the physical environment

Assessment of the physical environment is conducted with several goals in mind. Perhaps the most important of these relate to the planner's need to understand how the physical environment exerts a controlling influence on the resource potential of land. From this perspective assessments are performed primarily to:

- Increase the efficiency of human investments.
- Minimize hazards to life and property.
- Protect water quality.
- Minimize soil erosion.
- Protect important aquifers and recharge areas.
- Preserve open space.
- Protect sensitive and unique natural areas.
- Identify potential development conflicts.

Because extensive knowledge of a site is essential before any proposed change can be entertained, assessment must emphasize the integration of natural factors into a definition of the total landscape. This holistic view enhances the identification of physical components and their interrelationships (Baldwin, 1985). While a detailed discussion of this topic is beyond the scope of this chapter, the major considerations that demand careful review can be examined, and the dominant control mechanisms, introduced by a selection of physical processes germane to the environmental planning approach, can be examined. From this general overview, the information needed to produce workable plans and to guide site-specific analysis can be better understood. We can begin our evaluation from the ground up, starting with the geologic environment.

Geologic controls

Geologic information provides the basis for understanding the physical processes that shape the planning area. From bedrock properties which form the fundamental definition of stability, to the landform building processes and the active agents of weathering, geology describes the stage on which development takes place. In the same context, the geologic environment also explains a set of properties that conspire to (1) constrain the scale and intensity of development proposals, (2) introduce hazard, and (3) produce limitations that restrict the feasibility of committing land to particular types of uses. For example, knowledge of rock types and the environments in which they have formed, as well as their responses to weathering, erosion, and tectonic activities, becomes critical to the task estimating site conditions (Johnson & DeGraff, 1988). If the underlying bedrock is unstable, should the addition of water increase its weathering, then certain land uses may not be suited to that location. Thus, knowledge of the regional geologic history of the planning area is of great value, since this information broadens interpretation of an area and the relationships that exist between features such as rock type, their physical properties, their structure and geographic distribution, and how these factors influence the appropriateness of a given development plan.

Rock type

Rock may be defined in one of two principal ways. Geologically, rock is a naturally occurring consoli-

dated or unconsolidated material composed of one or more minerals. However, this definition also includes materials with physical properties that an engineer might consider to be soil. Therefore, engineering definitions of rock characterize it as a hard, compact, naturally occurring aggregate of minerals. In general the engineering properties of rocks are uniquely related to rock type (Johnson & DeGraff, 1988). This observation is particularly true when specimens are unbroken and intact. For this reason the name of the rock should provide information useful for engineering applications. For instance, in the example of most igneous and metamorphic rocks, mineralogy, texture, crystal size and structure are implicit in the name, as are the prevailing conditions at the time of the rock's origin. Not surprisingly, classification schemes for these rock groups are based logically on these variables. Sedimentary rocks, however, require more careful interpretation. The complex interaction of sedimentary environment, parent material, the detrital and/or soluble products of weathering, transporting mechanisms, lithification, and postdepositional changes preclude simple classification (Johnson & DeGraff, 1988). This fact carries important implications when one is estimating engineering properties based on the names and descriptions of sedimentary rock types.

Properties of igneous rock

Igneous rocks have silicate mineral compositions and interlocking textures (Johnson & DeGraff, 1988). Engineering classification of these rock types is based primarily on composition of crystal size. For comparatively recent igneous rock, mineralogy and texture combine to produce high strength and excellent elastic deformation characteristics. Thus, crystal size inversely affects strength. The emplacement mode of intrusive igneous rocks also has engineering significance, as does the origin of igneous extrusives. Knowing the boundary limit and rock type provides information on a variety of physical properties that may affect construction activities and slope stability, as well as the use of certain rocks as construction material.

Sedimentary rocks

Sedimentary rocks present the greatest challenges to planners and engineers. Because sedimentary rocks are the products of numerous marine, fresh-water, and terrestrial environments, they exhibit a wide range of physical properties, lateral extents, and thicknesses. In general, attempts to classify sedimentary rock are complicated by the fact that grain size separates the rocks composed of detrital material, while composition separates rocks of chemical and organic origin. An important characteristic of all sedimentary rocks is stratification. Primary sedimentary structures such as bedding surfaces and cross bedding create discontinuities in addition to those formed by secondary structures such as joints and faults. Primary and secondary structures reduce rock-mass strength and may contribute to slope instability (Johnson & DeGraff, 1988). The physical properties of sandstone, shale, and limestone are also influenced by differences in compaction, composition, grain size range, texture, and the nature and amount of cementing material. Vertical and horizontal gradation into other sedimentary rock types may also be expected.

Metamorphic rocks

Metamorphism causes textural, structural, and mineralogical changes in the original rock, modifying its physical properties (Johnson & DeGraff, 1988). These modification may improve some engineering properties, while others may contribute to reductions in strength, slope stability, and abrasion resistance. Metamorphic rocks are classified primarily on the presence or absence of foliation. The significance of this attribute of metamorphic rock is that foliation degrades a rock's engineering properties.

Planning significance

The key to using geologic information effectively depends on the planner's ability to "see" where this information fits. The previous discussion of rock and rock types is important simply because most construction takes place on or into rock.

Therefore, the engineering characteristics of the underlying rock will influence where specific structures are placed, how they are designed, and the methods used to build and maintain them. Any project involving the addition of great weight on the surface requires a detailed understanding of how rock will react under a variety of conditions. The planner must therefore obtain answers to several very basic questions pertaining to how the rocks involved will react when they are wet or dry, when they freeze and then thaw, when the dip of the rock is flat, gentle, or steep, or when they are subjected to earthquakes or large-scale subsidence. Since certain characteristics of rocks govern their stability, factors such as mineral composition, degree of weathering, the structure of the rock mass, the porosity and permeability of the rock, the depth to bedrock, and the depth to aquifer must be known. Each of these factors greatly influences the ease of excavation, foundation support, and other critical construction properties that will determine whether a building design, type, or placement is appropriate for a given site.

Weathering processes

Weathering defines a series of physical and chemical processes that are responsible for producing sediments that may lithify into sedimentary rocks or occur as engineering soils. When one is considering the appropriateness for a given development plan, perhaps the most important factor is the role weathering plays in altering the engineering properties of rock and rock masses. As an active process, weathering will influence all rock types, depending on their resistance, the type of weathering process involved, the environment to which the rock is subjected, the local climate, and time. Weathering processes can be either physical or chemical.

Physical weathering describes the mechanical breakdown of rock as the result of thermal expansion and contraction, unloading, hydration, and swelling. The most significant product of physical weathering are talus slopes and rock falls that originate from frost wedging and gravity movement of jointed rock masses. These features contribute to construction and maintenance problems and identify hazards to structures and facilities placed in proximity to locations where such processes are active.

Chemical weathering explains the decomposition of rock by chemical reactions that alter its composition. Chemical weathering processes include oxidation, solution, and hydrolysis. Of these agents, solution and hydrolysis are perhaps the most significant to the planner and engineer (Johnson & DeGraff, 1988). These forms of chemical weathering contribute to the weakening of rock and its gradual disintegration. Where active, these processes can produce massive failures at the surface, leading to the formation of sinkholes where underground support has given way. For example, the widening of joints in limestone and the development of caves are good examples of how solution alters subsurface properties.

Geomorphic controls

Geomorphology can be defined as the study of landforms. To the planner or engineer, the origin and nature of the landform is a key indicator of the processes that actively shape the land surface. These processes also provide important clues as to how the present land surface was formed and which geologic controls exert an influence on the location and scale of development. When examining the geomorphic characteristics of the landscape it is valuable to the planner to recognize that when observing the land surface, there exists a delicate balance or equilibrium between landforms and process, and the character of this balance is revealed by considering both factors as systems or parts of systems. In addition, the perceived balance between process and form is created by the interaction of energy, force, and resistance. Therefore, changes in driving forces and/or resistance may stress the system beyond the defined limits of stability. Furthermore, various processes are linked in such a way that the effect of one process may initiate the action of another.

With these conditions understood, we can see that the landscape explains a pattern where erosion and transport agents combine with weather-

ing processes to create characteristic forms such as valleys, ridges, fans, moraines, dunes, and so forth. The principal agents responsible for these unique forms are water, ice, wind, and mass wasting. Thus, insight regarding the geologic structure of the surface can be gained through the careful examination of landforms and the drainage patterns over them. For the planner, a set of questions can be posed with respect to the geomorphology of the planning area that help improve knowledge of form and process and clarify their planning implications (after Cooke & Doorkamp, 1990):

- Which forces define the present geomorphic situation, and underscore its present-day problems, hazards, and resources?
- What is the nature and rate of geomorphic change at present, and how stable is this landscape?
- What will happen to geomorphic processes if the area is developed?

We can examine a sample of geomorphic agents to help shed light on these questions.

The role of drainage

One of the more useful indicators of surface and subsurface characteristics under similar conditions of climate is drainage density and pattern. For example, when viewed from a map or aerial photo, lower density or more widely spaced drainage channels indicate subsurface drainage through thick, permeable soils, or through karstic limestone terrain when compared with greater drainage density or less permeable soils and bedrock units (Johnson & DeGraff, 1988). Drainage patterns also tend to help define rock types and geologic structure (Way, 1978).

Typically, drainage patterns are classified either by their density and dissection or texture, or by the type of pattern form they exhibit. Drainage texture is defined in three principal ways:

1. *Fine-textured* – indicates high levels of surface runoff, impervious bedrock, and soils of low permeability.
2. *Medium-textured* – describes conditions where the spacing of first-order streams is less than fine textured, and the amount of runoff is medium when compared to that associated with fine or coarse textures

Dendric   Rectangular   Trelis

Dichotomic   Radial   Parallel

Centripetal   Braided   Artificial

Fig. 4.2 Characteristic drainage types.

3 *Coarse-textured* – indicates a more resistant bedrock which may be permeable and forms coarse, permeable soils.

In addition to drainage texture, drainage pattern is also a useful indicator of landform structure and process. There are 23 major descriptions of drainage types. These categorizations encompass the entire pattern of water flow, including gullies or first-order area of channelized flow, the tributaries, and the major channels which may be depositing eroded materials, to form surficial water-laid landforms. The main patterns classified according to this scheme are illustrated in Fig. 4.2. Each of the patterns shown can indicate the presence of specific rock types, soil materials, rock structures, and drainage conditions. An excellent

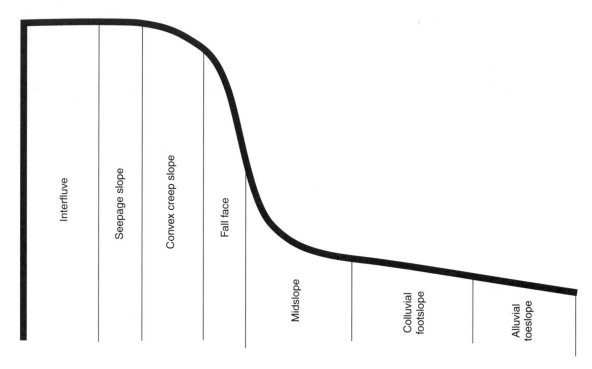

Fig. 4.3 Generalized slope characteristics.

description of the planning significance of the drainage patterns illustrated in Fig. 4.2 can be found in Way (1978).

When the planner is assessing the geologic environment, his or her search for information revolves around a series of questions that not only define critical information needs but also point to the geologic factors that should be used to form an inventory of the planning area (Flawn, 1970):

- What part of the planning area is most suited for the desired future land-use need?
- Where in the planning area are excavation costs high and where are they low?
- Where in the planning area are there highly corrosive earth materials?
- Where in the planning area do earth surface processes represent risk or hazards?

Topographic controls

Topography forms the basic definition of the form and character of the terrain. Considered synony-

mous with the concept of relief, topography defines the overall configuration of the landscape. Perhaps the most important expression of topography to the environmental planner is slope (Fig. 4.3). Aside from its influence on the use and capability of land, slope exerts substantial controls on runoff, ground stability, and erosion. Slope, defining the inclination of a landform from the horizontal, can be expressed in several ways:

- Steepness of slope – defined by the vertical gradient as interpreted from the contours of a topographic map.
- Length of slope – the physical distance from the head to the toe of the landform.
- Percent of slope – the ratio of slope defined as a function of elevation over distance expressed as a percentage.
- Slope gradient – the maximum rate of change in altitude.
- Slope angle – the angle of inclination from the horizontal.
- Slope form – the morphology or shape of the slope.

- Slope composition – the earth and rock material that constitutes the landform.
- Slope stability – the inherent resistance of the slope to failure.

The need to consider slope in planning is the direct result of the realization that slope not only imposes limitations on the use of land, but that modern development practices have contributed to a misuse of sloping land. Misuse of slope occurs in two fundamental ways (Marsh, 1997):

1 The placement of structures and facilities on slopes that are unstable or potentially unstable.
2 The disturbance of stable slopes resulting in failure, accelerated erosion, and ecological deterioration of the slope environment.

There are three common disturbances that contribute to the majority of slope-related problems. These include (1) mechanical cut and fill where the gradient and length of slope is reshaped by construction, (2) deforestation, where in hilly terrain urbanization, agriculture, and extractive industries encourage the removal of vegetation that alters the hydrologic regime and stability of slopes, and (3) drainage alterations resulting from the construction of buildings on slopes which change slope equilibrium and the surface and subsurface flow of water.

Slopes are among the most common landform and although most appear to be stable and static, they should be viewed as extremely dynamic and evolving systems (Keller, 1988). Because of their dynamic nature, slope stability is a major concern in planning. There are a number of ways in which sloped earth materials may fail and move or deform. Several of the more pronounced are flowage, sliding, falling, and subsidence. The variables that influence downslope movement include factors such as the type of slope movement, the material composing the slope, the rate of rock movement, and the amount of water present on the slope (Keller, 1988).

Slope stability

Depending upon the type of movement – creeping, falling, sliding, or flowing – and the kind of material involved – rock, earth, or mud – mass movements can vary considerably in their shape, rate, extent, and effect on surrounding areas (Griggs & Gilchrist, 1977). Understanding slope stability issues is paramount for effective planning. In general, the nature and extent of slope stability can be explained by examining forces on slopes in relation to 6 interrelated factors:

1 type of earth material
2 slope characteristics
3 climate
4 vegetation
5 water
6 time.

Slope forces

The propensity for downslope movement is defined by the interplay between a set of driving forces that tend to move earth materials down a slope and the set of resisting forces that act to oppose downslope failure. The most common driving force is the downslope component of the weight of the slope material. This factor includes anything superimposed on the slope, such as vegetation, fill material, or buildings (Keller, 1988). The most common resisting force is the shear strength of the slope material acting along potential slip planes. According to Griggs and Gilchrist, the shear strength of a material is defined as the maximum resistance to failure. This quantity can be described in relation to:

1 Internal friction due to the interlocking of granular particles, and
2 Cohesion due to the forces that tend to hold particles together in a solid mass.

Thus, as slopes become steeper, the shear stress (defining the force of gravity lying parallel to a potential or actual slippage) exerted on surface material increases as the downslope pull of gravity increases. A number of internal and external factors can affect the balance between these forces. External factors affect the stress acting on a slope, while internal factors act to alter the strength of slope material. Critical external factors include:

- Change of slope gradient – where an alteration in the angle or degree of slope may degrade internal resistance to downslope forces.

- Excess loading – where the addition of weight at the head of a slope can destabilize internal equilibrium conditions.
- Change in vegetative cover – where the removal of vegetation can effect the overland and subsurface flow of water.
- Shocks or vibrations – where energy waves can effectively shake slope material.

Important internal factors that require careful review include:

- Change in water content.
- Groundwater flow.
- Weathering.
- Slope angle.

Characterizing slope

Slopes may be characterized in a number of ways, and a variety of graphical devices have been developed to help visualize slope information. Perhaps the most basic method to visualize slope is by means of a slope profile. A slope profile is constructed from the contour lines displayed on topographic maps. Interpretation of the profile permits a general categorization of slope form (Marsh, 1997). A more useful device for assessing slope and making judgments regarding slope characteristics is the creation of a slope map. The purpose of such a map is to graphically indicate each slope and place slope characteristics into a classification system to order and highlight their significance. The general sequence of steps needed to manually produce a slope map using a topographic map can be outlined as follows:

1 Determine the slope classes that will be used to order slopes, e.g., 0–10%, 11–25%, >26%.
2 Note the contour interval and scale of the base map, e.g., 40-foot interval, with a scale of 1 inch to 2,000 feet (1:24,000).
3 On one edge of an index card or similar durable medium, carefully inscribe five tick marks at 400-foot intervals (note that 40/400 =10%). Identify this scale as 10%.
4 On another edge of the index card inscribe another series of ticks at 160-foot intervals (40/160=25%). Identify this scale as 25%.
5 Only these two classes need to be marked on the card, since any area where the contours

are closer together than 160 feet would be over 25%.

6 Using the base map delineating the planning area, locate areas of 10% slope by placing the appropriate card scale at right angles to the contour. At each location where the contour lines and scale marks coincide, place a mark on the base map. If the distance between contours is greater than 400 feet (each index mark), placing a mark on the base map is unnecessary, since this area would be less that 10% and still fall within the same slope class. By connecting the marks placed on the base map and the perimeter contour, the outline of areas with slopes between 0 and 10% can be delimited.
7 Following the same procedure explained above in Step 6 for the remaining slope classes completes the preparation of the slope map.
8 Each of the classes mapped should be symbolized according to a color or shading scheme that permits sound cartographic representation of the slope information and enhances visual interpretation of slope data.

The manual method of slope mapping using index cards tends to be a tedious and labor-intensive process. Fortunately much of the drudgery has been taken out of slope mapping by the widespread availability of digital elevation models (DEMs) and use of specialized routines found within the toolbox of leading geographic information systems (GIS) software. Whether hand drawn or produced via a GIS, slope maps are an important way of communicating critical information regarding the nature of slope in the planning area.

Soil considerations in planning

Soils describe a transitional environment juxtaposed between the lithosphere and the biosphere/atmosphere complex. Although soils are frequently overlooked as a major factor in planning analysis, their influence on the design and sustainability of land uses is noteworthy. One possible reason that may account for the lack of concern for soils in planning has to do with the contrasting definitions used to explain soils and soil characteristics. Soils are the natural bodies in which plants grow. They underlie the foundations of houses, factories, and commercial establishments, and determine whether these foundations are adequate. They serve as the beds upon which roads and highways are constructed and influence the life expectancy of these features. Soils are also used to absorb and filter waste from sewage systems. Therefore their importance is indisputable, although a simple definition of what they are may remain elusive.

Soil refers to the loose surface of the earth and can be explained as: (1) a medium for plant growth, (2) a mantle of weathered rock, or (3) an engineering material. Under these differing conceptual views, soil may be defined simply as unconsolidated mineral matter on the surface of the earth that has been subjected to and influenced by a series of genetic and environmental factors. The dominant influences on soil are

- Parent material – the bedrock properties from which soils are derived.
- Climate – the thermal and precipitation regimes that influence weathering and biological processes.
- Macro- and micro-organisms – which assist in modifying soil composition.
- Topography – characterizing slope and indirectly influencing rates of erosion.

Each factor, acting over time, produces a soil that differs from the material from which it was derived in several ways, including its physical, chemical, and biological properties and characteristics (Fig. 4.4).

To the planner, the physical properties of soil greatly influence its suitability with respect to the many land uses that may be placed on top of or into its structure. In addition, the rigidity and supporting power, drainage and moisture-storing capacity, plasticity, ease of penetration by roots, aeration and retention of plant nutrients are all intimately connected to its physical condition (Foth, 1978). Several of the more salient properties of soil with particular relevance to the land development problem include:

- **Soil composition** – refers to the materials that make up a soil. There are four primary

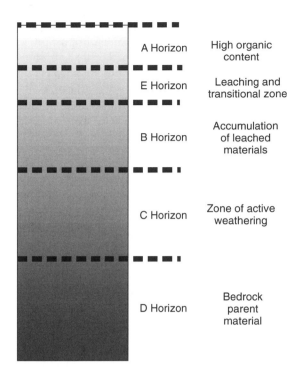

Fig. 4.4 A typical soil horizon.

constituents of a soil: mineral particles, organic matter, water, and air. Mineral particles comprise 50% to 80% of the volume of most soils. Mineral structure allows the soil to support its own weight as well as that of water and buildings placed overlying the soil. A soil's mineral composition helps to define its bearing capacity, a quantity that describes a soil's resistance to penetration from a weighted object. Organic composition can vary widely in soils. Generally, organic particles provide a weak structure with poor bearing capacities. Thus, soils with high organic content tend to compress and settle differentially under road beds and foundations. Furthermore, when dewatered, these soils may undergo substantial volume losses through decomposition and wind erosion.

- **Soil texture** – refers to the fineness or coarseness of the soil expressed relative to particle size. In more specific terms, texture characterizes the relative proportions of sand, silt, and clay. The rate and extent of many physical and chemical reactions important to plant growth are governed by texture, since it determines the amount of surface on which these reactions can occur (Foth, 1978).

- **Soil structure** – refers to the aggregation of its primary particles (sand, silt, clay) into compound particles which are separated from adjoining aggregates by surfaces of weakness. Therefore, the structure of the different horizons of a soil profile is an important characteristic of the soil, as are its color, texture, and chemical composition. Structure tends to modify the influence of texture relative to the presence of moisture and air, the availability of plant nutrients, and the actions of micro-organisms and root growth.

- **Soil moisture and drainage** – A soil's water content will vary with particle size, topography, texture, and climate. Water can occur in two principal ways in either mineral or organic soils:

  1) Capillary water – a form of molecular water held in the soil by cohesion among water molecules.
  2) Gravity water – characterizing liquid water moving in response to gravitational forces. This downward flow accumulates in the subsoil and underlying bedrock to become groundwater.

Drainage properties of soil typically related to gravity, water, and the capacity of a soil to transmit water downward. Three important factors help define soil drainage and also serve as critical determinants of land-use suitability:

1 Infiltration capacity – which refers to the rate at which water penetrates the soil surface. Infiltration capacity is greatly influenced by the structural stability of the upper horizons of a soil. Other factors such as organic matter content, soil texture, soil depth, and the presence of impervious soil layers help to refine the infiltration capacity rate.
2 Permeability – refers to the rate at which water transfers through a given volume of material. The permeability rate of a soil defines the ability of the soil to transmit water. Permeability can influence decisions relating to the location and design of septic-tank systems, the size of terrace ridges, and the slope of terrace channels for erosion control, as well as the suitability of land for agricultural uses.
3 Percolation – explains the rate at which water in a soil pit or pipe placed into the soil is taken up by the soil. Percolation results in leaching, which can deplete a soil of certain nutrients. Typically, when the amount of precipitation entering a soil exceeds the water-holding capacity of the soil, losses by percolation will occur. Percolation losses in a soil are affected by the amount of precipitation, its distribution, runoff, and evaporation, and the overall character of the soil.

Other important qualities of soil relevant to the environmental planning problem include:

1 Corrosivity, which varies with soil acidity, texture, drainage, electrical conductivity, and the presence of corrosive agents such as sodium or magnesium sulfate.
2 Shrink–swell potential, which defines the interlayer expansion of silicate clays, particularly of the semetic and vermiculite units, given changes in water content.
3 Depth to bedrock, a characteristic describing the distance through the soil mantel before soil bedrock is encountered.
4 Heightened erodability, which describes the ease with which soils may be detached and moved by water, wind, ice, or gravity.

Each of these soil variables introduces limitations that restrict or redirect the use of land and determine its relative fitness for development.

Soil-related planning issues

There are a number of concerns that the planner must factor into the decision-making process when interpreting soil information and reviewing sites for various development proposals. For example, organic soils are highly compressible under the weight of structures and tend to decompose when drained. In the case of mineral soils, texture and drainage are important factors. Generally, coarse-textured soils present the fewest limitations; however, as clay content increases, drainage and shrink–swell potential pose constraints for development. Perhaps the most pressing concern in planning with soils is the issue of erosion. Soil erosion is both a natural process and one accelerated by human activity. As an active process, the wearing away of the land surface by water, wind, ice, or poor land management practices can assume many different forms. A series of the descriptions used to characterize different types of erosion has been offered by Foth (1978), and includes:

- Accelerated erosion – erosion much more rapid than normal natural geologic erosion, primarily as a result of the influence of people or animals.
- Geologic erosion – the normal or natural erosion caused by geological processes acting over long geologic periods and resulting in the wearing away of landforms and the build-up of flood plains and fans.
- Gully erosion – the erosion process whereby water accumulates in narrow channels and over short time periods, and removes the soil from these narrow areas to considerable depths.
- Natural erosion – wearing away of the earth's surface by water, ice, or other natural agents under natural environmental conditions of climate and vegetation, undisturbed by humans.
- Normal erosion – the gradual erosion of land used by people that does not greatly exceed natural rates.
- Rill erosion – an erosion process in which numerous small channels of only several inches in depth are formed.
- Sheet erosion – the removal of a fairly uniform layer of soil from the land surface by the overland flow of water.
- Splash erosion – the spattering of small soil particles caused by the impact of raindrops on very wet soils.

Soil erosion remains among the most significant problems in land management and environmental planning. Because virtually all land uses contribute directly or indirectly to the process, erosion is pervasive and results in direct monetary losses as well as a range of indirect costs or effects,

Table 4.2 Erosion control strategies.

Site selection practices
Reducing soil exposure
Structural control methods
Land-use management

such as (1) increased sedimentation and clogging of streams and channels, (2) heightened flood frequency and risk, (3) land degradation, and (4) reduced water quality. These considerations place a premium on developing methods to control and reduce the soil erosion risk.

Erosion control

Erosion can be controlled effectively if (1) study and planning precede any kind of land surface alteration, and (2) a set of basic principles are followed. Several methods have been shown to be effective, particularly with respect to urban development. A selection of these control strategies is listed in Table 4.2. The methods outlined are based on three guiding principles that can be applied to any development proposal. First, ensuring that environmentally appropriate sites have been selected for the proposed development. Secondly, taking steps to reduce the area and duration of exposure of soil to erosion. Lastly, adopting measures to mechanically retard runoff, or trapping the sediment removed from a site. Following these recommendations and integrating them into development plans can reduce the adverse effects typically associated with soil erosion.

Climatic and hydrologic considerations

Climate and planning

The effects of land use and land cover on local climate have been well documented (Lein, 1989). The question for the planner to consider is not simply the relationship between land development and climate, but the larger issue of how climate can introduce opportunities for and constraints on sustainable development. To understand climate's role in sustainable planning one need only begin with the influence climate has on thermal patterns, air pollution dispersion, energy efficiency, and human comfort, and the ways in which these factors can be modified by the form and material structure of the land surface. Gaining this appreciation for climate in planning begins with a review of the fundamental principles that determine and control local climate, and the properties of the climate the stand to be modified by changes at the surface.

Climate can be defined as the aggregate of atmospheric conditions involving heat, moisture, air movement, and their extremes, measured over periods of months to years (Henderson-Sellers & Robinson, 1986; Hidore & Oliver, 1993). Of the various scales by which climate may be examined, two of the more pertinent to the environmental planner are (1) the synoptic or mesoscale and (2) the microscale. Climate at the synoptic scale describes the regional patterns and processes that form unique characteristics associated with distinctive geographic features such as mountains or valleys and explain regional controls that dominate or influence atmospheric conditions. Here, the seasonal fluctuations of major pressure systems and the physical controls of climate within the mesoscale environment exert influences that give rise to distinctive climatic regimes such as lake-effect characteristics, thermal inversions, summer drought patterns, and convective circulation complexes (intense thunderstorm cells). Local, microscale climates are defined by environments where specific topographic and land-cover arrangements and configurations exert moderating influences on the pattern established by the dominating synoptic controls. Examples of these small-scale patterns include forest climates, urban climates, and climates that are topographically induced.

Regardless of scale, an important means of characterizing climate is through the mechanisms that direct the exchange and budgeting of heat energy between the surface and the atmosphere. Here, the input of solar radiation that drives the climate system is subject to a series of transfer mechanisms that redistribute this flux and in the process produce fundamental climatic regimes.

This budgeting of solar radiation is also a very useful way to explore the effects of surface change on the disposition of climate at global to local scales. The flux of energy at the surface can be expressed symbolically by the relation

$$R_n = (Q + q)(1 - \alpha) + I\!\downarrow - I\!\uparrow \qquad (4.1)$$

Where:

R_n = net radiation
Q = direct solar radiation
q = diffuse solar radiation
α = surface albedo
$I\!\downarrow$ = atmospheric counter-radiation (long wave)
$I\!\uparrow$ = terrestrial radiation (long wave).

As suggested by the above relationship, the concept of balance is critical to understanding the climate system. Because the only energy input to the system is solar radiation, this extraterrestrial short-wave flux entering the earth atmosphere system must, on an annual basis, be balanced by an equal amount of energy leaving the system. Thus, as shown in equation 4.1, the radiant energy available to drive climate (R_n) is the product of the energy reaching the surface directly or after diffusion by clouds in the atmosphere ($Q+q$), reflection off the surface ($1 - \alpha$), and transfer from the surface to the atmosphere and from the atmosphere back to the surface. Any change in these terms will change the value of R_n and alter climate. The implications are subtle, but they form the basis for understanding human-induced climate change. Therefore, promoting human activities that may alter the composition of the atmosphere may result in changes in the quantity and quality of radiation reaching the surface. Once at the surface, land cover can redirect the quantity of radiation reflected off the surface. Changes in the quantity of radiation at the surface will change the rate of terrestrial radiation, and so on. If atmospheric scattering and absorption are included along with surface and atmospheric storage, the fundamental balance relationship can be expanded to produce a more complete description:

$$R_n = R_g(1 - \alpha) - \Psi_c - \Psi_a + \beta_c + \beta_a + \Delta_s + I\!\downarrow - I\!\uparrow$$
$$(4.2)$$

Where:

R_g = global radiation
α = albedo
X_c = reflection by clouds
X_a = reflection by aerosols
β_c = absorption by clouds
β_a = absorption by aerosols
Δ_s = absorption by the surface
$I\downarrow$ = atmospheric counter-radiation
$I\uparrow$ = terrestrial radiation.

Given the net radiation input at the surface, the balance of energy is achieved for a specific surface by a series of transfer functions that define the energy conservation equation for the surface:

$$R_n = H + LE + \Delta G \qquad (4.3)$$

Where:

H = sensible heat flux
LE = latent heat flux
G = surface heat flux.

The significance of the relationships expressed above it that they identify the processes involved in forming climate at scales of interest to the environmental planner. They also suggest pathways whereby changes at the surface can modify climate processes and alter surface conditions. For example, changing surface vegetation and exposed soil to asphalt and concrete will change surface albedo, alter the flux of latent heat by removing vegetation, and increase sensible heat flux from the surface. Each of these deviations from the original conditions presents new microclimates and can contribute to a redistribution of solar radiation. While such microclimatic alterations appear to be subtle and insignificant, they have been shown to carry important implications in site planning and design (Douglas, 1983). These include:

- The way in which the walls and roofs of buildings and concrete paved surfaces behave like exposed rock materials and display high heat capacities, conductivities, and evidence an ability to store and reflect heat energy at rates greater than that for natural surfaces.
- The way in which the additional surface areas of buildings with large vertical faces

create exchanges of energy, mass, and momentum.

- The input of artificial heat generated by machinery, vehicles, and heating and cooling systems.
- The way in which a large extent of impervious surface removes precipitation quickly and alters moisture and heat balances by reducing evaporative processes.
- The ways in which the injection of pollutants and dust into the atmosphere modifies radiation and energy-balance processes.

Based on these examples, as the surface is transformed from a natural to a functional state, development modifies the ambient energy balance, air circulation, and evaporative processes through the introduction of multiple reflection and absorption patterns, rough and uneven surface forms, and by altering vegetative cover. Modification also results from the addition of sources of heat, dust, and pollutants that intensify as the pace of human activity increases. The significance of these changes will vary with prevailing synoptic scale controls, yet the ground-level consequences to human comfort, human health, and energy efficiency have been well documented. The examples presented above also suggest that the natural advantages or amenities associated with climate that guide siting and design may be degraded as the intensity, scale, and pattern of development promotes continued modification.

Hydrologic factors

The process description used to explain climate's role in planning is also a convenient way to characterize the hydrologic regime and how it influences the planning process. Consideration of the hydrologic regime directs attention to the variables that define the hydrologic cycle. Perhaps the most comprehensive treatment of hydrologic cycle and related processes with specific reference to environmental planning can be found in Dunne and Leopold (1978).

The hydrologic cycle can be represented symbolically as

$$P = E \pm T \pm R \pm I \pm \Delta S \qquad (4.4)$$

Where:

P = precipitation

E = evaporation

T = transpiration

R = runoff

I = infiltration

S = soil-moisture storage.

According to this relation, the input of precipitation at the surface is balanced by a series of output or transfer mechanisms that redistribute water back through the earth/atmosphere system. In a manner similar in concept to the surface-energy balance, any change in the variables on the right-hand side of the equation will alter the availability and movement of water within the planning area. However, as built form replaces natural landscape, the movement of water is actually influenced by two interrelated hydrologic systems: (1) the human-modified hydrologic cycle and (2) the human-created water supply and water removal system (Douglas, 1983). These contrasting expressions of water, when coupled with the removal of vegetation, the addition of impervious surfaces, the alteration of soil qualities, the modification of drainage, and the utilization of both surface and subsurface water introduced by land development practices, generate a series of hydrologic effects such as:

1 A change in total runoff at the surface.

2 An alteration of peak flow characteristics.

3 A decline in the quality of water.

4 Changes in the hydrologic amenities associated with rivers, lakes, and streams.

Add to these documented effects changes introduced by the network of water collecting, treating, transmitting, and distributing systems created to provide for development. With these active systems, water is abstracted, relocated, and discharged in a manner that can upset the total hydrologic balance of a watershed (Douglas, 1983). Therefore, recognizing the hydrologic implications of land use and land-cover change on the balance, inflow, transfer, and outflow of water is an integral step in planning sustainable urban systems. However, with *any* alteration of the surface, a change in the natural hydrologic cycles can

be anticipated. The most noteworthy surface changes include the clearing of forested lands, the overgrazing of land, and the introduction of impervious surfaces.

As suggested by the processes involved, as the composition of the surface changes, the rates of evaporation, transpiration, infiltration, and runoff will deviate from their expected condition. Through the transformation of land from natural to urban use the planner can anticipate:

- Decreased infiltration – with the addition of impermeable surface, less ground area is available to allow water to percolate into the soil.
- Increased runoff – with increases in impermeable surface, overland flow will accelerate.
- Accelerated erosion – with faster rates of overland flow the erosion potential of the surface will increase.
- Altered flood regimes – with changes in infiltration, runoff, and erosion, sedimentation and flooding will be more common and contribute to urban flood problems and ponding.

Although these concerns are often viewed individually, they are highly interrelated and serve to illustrate the importance of examining both direct and indirect effects as land-cover changes are contemplated.

Biotic and ecological considerations

The ecological approach to planning is predicated on a detailed understanding of ecological process and the interrelationships that exist between geology, soils, climate, and hydrology and the structure and function of the ecosystems that define the planning area. Although the ecological approach to planning is widely advocated, distilling the essential knowledge of ecosystem processes required to produce ecologically sound plans remains a challenge (Steiner, 1991; Park, 1980; Peck, 1998; Sukopp et al., 1995). Much of the problem associated with ecological planning stems from the fact that urban landscape design continues to operate on the premise that ecological processes are either non-existent in cities or have little relevance to the issues of design and form (Hough, 1995). Ecological processes, however, provide an indispensable basis for planning if the planner recognizes the significance of basic ecological principles. Fundamental ideas such as the dependence of one life process on another; the interconnected development of living and physical processes on the earth; the continuous transformation and recycling of living and nonliving materials, define elements of a self-perpetuating biosphere that sustains life and directs the form of the physical landscape. These same elements become central determinants of the form underlying all human activities as well.

Ecosystem planning

The ecosystem has become the fundamental unit of study in ecology and represents a concept that is of great value to all aspects of environmental planning. It is here, at the ecosystem level, where the ecological approach begins. For our purposes, an ecosystem can be defined simply as a self-regulating association of plants and animals interacting with one another and their nonliving environment in a well-defined geographic area. Self-regulation is a key concept in this definition, and serves as the starting-point for achieving sustainable development goals. Through the arrangement of the various living elements of the system, the relationships they assume together with the nonliving elements that characterize the physical and chemical factors that form an environment, a structure, pattern, and flow is established which moves matter and energy through the system. This idea, while rudimentary, is all too easy to overlook as development pressures encourage change at the surface.

The components that comprise an ecosystem include

1 Producers – defining plants that convert sunlight into chemical energy through photosynthesis.
2 Consumers – describing animals that feed directly or indirectly on producers, which can be further divided into more specialized

groups that explain their particular role or function:
- Herbivores – animals that feed on plant material.
- Carnivores – animals that feed on other animals.
- Omnivores – animals that feed on either plants or animals.
- Decomposers – organisms that break down waste plant and animal material and return nutrients to the soil.

When diagrammatically represented, these components form a fundamental system structure that can be used to illustrate key functional relationships. The defining elements of an ecosystem are connected by a flow of energy as the input of solar radiation is transformed into a new energy source that enters the food chain and cycles through the system. Other elements cycle through the system as well. These nutrient cycles include the carbon, nitrogen, and phosphorus cycles, and serve to reinforce the concept of self-regulation and support the dynamic balance ecosystems achieve over time.

Our review of basic ecology carries several important implications and points to the information needs the planner will require to effectively balance development with sustainable environmental functioning. First, the functioning of an ecosystem reveals that the plants and animals defining its structure participate in a complex flow of energy that supports and perpetuates the system. To facilitate this flow each organism has a niche that defines all the physical, chemical, and biological factors that it needs to survive and reproduce. This niche includes considerations such as an organism's:
- Food niche – the specific foods that compose a species' diet.
- Habitat niche – the environment that the species inhabits.
- Reproductive niche – the pattern and characteristics of species' ability to regenerate.
- Physical and chemical niche – environmental elements required to sustain the species.

Plans and development proposals will alter niche relationships that can affect whether or not a given organism can find the food it requires and the habitat it needs, mate successfully, and acquire the other factors it needs to survive. These effects can be very subtle and reveal their effects after long periods of time have elapsed, or they can result from major and immediate disruptions and redirections in the quality and quantity of energy flowing through the components in the system. In either case this pattern introduces a disturbance which affects the stability of the ecosystem and ultimately degrades its resource potential. At this point in out brief discussion it must be recognized that natural ecosystems, due to their many interdependencies and interrelationships, are invariably richer in species and more stable that those artificially developed. The urban pattern superimposed onto the natural landscape defines an ecosystem contrived by people. All of the facilities and arrangements of its parts are for human use. The material and energy flow arrives from well beyond the boundaries of a given urban complex. Furthermore, being "artificial," interdependencies are missing and the system is not self-regulating or self-sustaining.

The contrast between these two systems, the natural and human, together with the disturbance development introduces as interrelationships are severed by the implacement of built forms, suggests that land development processes will change the biotic community. However, because every environment has some recovery potential and level of adaptability, the effects on the natural system will vary in terms of intensity and duration.

The complexity and variability of ecosystems and their response to disturbance is further complicated by the fact that planning focuses on landscape and not on ecosystems. This distinction is important, since the landscape is almost always highly heterogeneous and can include such diverse elements as fields, woods, marshes, towns, and corridors (Gordon & Forman, 1983). For example, in a suburban area, a cluster of landscape elements such as a residential tract, a patch of woods, a commercial area, and an open, grassy field would define a pattern replicated throughout the landscape, whereas a different cluster of ecosystems would be found in adjoining agricul-

tural fields, or between a sandy, forested plain and a riparian corridor (Gordon & Forman, 1983). The landscape is therefore an area where a cluster of interacting ecosystems is repeated in similar form. Consequently, basic ecological principles pertaining to energy flow, nutrient cycles, niche concepts, balance, and successional dynamics of ecosystems provide a useful backdrop for a more landscape-directed focus. This focus is provided by the developing science of landscape ecology.

Landscape ecological planning

The application of landscape ecology theory has had a pronounced influence on environmental planning (Foreman, 1998; Foreman & Gordon, 1986), and there is growing interest in using landscape ecology as a scientific base for land-use decision-making (Hersperger, 1994). Two reasons may be cited to account for its appeal: (1) landscape ecology gives emphasis to the interface between humans and environment, and (2) landscape ecology recognizes the importance of change as a fundamental feature of the landscape. A landscape-ecological approach to environmental planning, therefore, tends to be more focused on the nature of spatial change and the functional connections between biophysical and sociocultural processes and agents of change.

According to the principles of landscape ecology, the landscape system shares three important characteristics: structure, function, and change (Hersperger, 1994; Foreman & Gordon, 1986). For the most part, structure describes the spatial relationship between the distinctive ecosystems or elements present in the landscape. The main landscape components that define a landscape's spatial structure are patch, corridor, and matrix. The overall landscape is therefore the synthesis of these basic parts, while the distribution of energy, materials, and species in relation to the size, shape, number, type, and arrangement of each differentiate the pattern. Accordingly, patch, corridor, and matrix can be explained as functions of factors such as origin, shape, size, connectivity, porosity, edge, heterogeneity, and configuration. Function defines the ecological processes at work in the landscape. Explaining function requires an understanding of how energy, water, mineral nutrients, plants, and animals move across landscapes composed of various combinations of structural features. Key factors used to define function include the relative structural attributes of the landscape, flow gradients (locomotion, diffusion, mass flow), and scale. Coupled to these qualities are the natural processes and human actions that influence landscape evolution on the larger scale. Because function influences structure and structure influences function, their interaction forms the logical basis for causal analysis. In this context, change is the product of the relationship between function and structure expressed over time. Change, therefore, implies temporal analysis and can be measured (or quantified) as an alteration in either structure or function. Subsequently, change, or its absence, is typically expressed in relation to stability. Thus, natural and human disturbances together with the natural processes that direct ecosystem development form the driving forces of change.

The concept of a disturbance and the methods developed to compare landscapes under differing levels of modification provide essential information to the planner: particularly with respect to issues related to conservation, biodiversity, and the development of sustainable land-use patterns. Generally, all landscapes are defined by a common geomorphic origin and a common disturbance regime. A disturbance regime describes the sum of types, frequencies, and intensities of disturbances through time. The disturbance itself is considered to be something that causes a community of ecosystem characteristics, such as species diversity, nutrient output, biomass, and vertical or horizontal structure, to exceed or drop below its homeostatic range of variation (Gordon & Foreman, 1983). Geomorphic processes such as uplift, erosion, aeolian action, and deposition produce physiographic units, which, in the absence of human activity, contain various natural disturbances. Landscapes modified by human activities are changed by new disturbances induced by humans. Examples include agricultural activities and urbanization. These human disturbances differ according to landscape and are superimposed onto natural disturbances and geomorphic settings, resulting in landscapes that vary widely and form distinct boundaries.

Comparing landscapes along a gradient from lesser to increasing levels of human modification can be accomplished by examining critical ecological attributes. These attributes have been summarized by Gordon and Foreman (1983) and include:

- *Horizontal structure* – In natural landscapes subjected to minimal human disturbance, horizontal structure shows little contrast. Primary breaks result from natural disturbances and from geomorphology. As human modification increases, sharp delineations between ecosystems become more numerous. This includes the proliferation of straight lines in the landscape, forming the borders between ecosystems and corridors crossing ecosystems. The dominant human imprint on structure involves: (1) increases in contrast, and (2) increases in linearization and rectangularization.

- *Stability* – Natural landscapes display considerable potential energy in the form of biomass, yet they are also stable. With increased modification, potential energy and stability decrease.

- *Thermodynamic characterization* – Although ecosystems vary in degree of thermodynamic openness, natural landscapes generally are more closed than modified landscapes. Modified landscapes receive a large external supply of nutrients and fossil energy. Human activity tends to increase openness of energy and nutrient cycles, which accounts for one of the principal ways by which human actions stress or modify the stability of the landscape.

- *Chorology* – describes the processes and results of species dispersal. In natural landscapes dispersion of reproductive structures is rather viscous. Managed forests and grasslands produce more fluid chorologic systems, while at the end of the modification gradient, the most human-influenced landscapes provide special opportunities for cosmopolitan species.

- *Minimal grain* – defines the minimal land area in which nearly all species of a community are found.
- *Net production* – characterizes the balance between photosynthesis and respiration. In natural landscapes the annual balance of plant production is close to zero. With increasing management, a positive net production is realized; however, in urban landscapes net production declines and can be negative.

Other critical factors to consider when examining disturbance patterns include nutrient cycling, tactics, phylogeny, and type of resistance.

Although landscape ecology offers theory and empirical evidence to help understand and compare different spatial configurations, few land-planning principles based on this knowledge have emerged (Foreman, 1998). One useful idea that has developed out of the landscape ecology school is the "aggregate-with-outliers" principle discussed by Foreman (1998). According to this principle, when considering the spatial allocation of land uses, care should be taken to:

1 aggregate land uses;
2 maintain corridors and small patches of nature throughout the developed area;
3 geographically arrange outliers of human activity along major boundaries.

From this simple beginning, several landscape qualities can be maintained and incorporated into plans as a basis for design. By so doing the overall functioning of the landscape as sustainable habitat can be enjoyed and a balance can be achieved between human use of the landscape and the natural processes at work there. A selection of qualities identified by Dramstad et al. (1996) include:

- **Large patches of natural vegetation:** are ecologically important because they: (1) protect aquifers, (2) protect low-order stream networks, (3) provide habitat for large home-range species, (4) support viable population sizes of interior species, (5) permit natural disturbance regimes in which species evolve and persist, and (6) maintain

a range of microhabitat proximities for multihabitat species.

- **Grain size:** The average area of all patches in the landscape affects many ecological factors. A landscape containing a variance in grain size (coarse and fine grain) is an important configuration.
- **Risk spreading:** Reducing the adverse consequences of limited diversity by encouraging a greater mix enhances sustainability.
- **Genetic variation:** Providing strains with resistance to disturbance and responding to environmental change by facilitating re-establishment.
- **Boundary zone:** Boundary zones between land uses are often suitable for outliers. When located along boundary zones, outliers do not perforate and destroy the advantages of large patches. The curvilinearity of boundaries also reduces barrier effects and mimics the results of natural processes.
- **Small patches of natural vegetation:** Small patches or outliers of natural vegetation are valuable throughout the developed area. Small patches are a significant supplement to large patches in that they: (1) serve as stepping-stones for species dispersal, (2) provide habitat and stepping-stones for species recolonization following local extinctions, (3) provide heterogeneity in the matrix that decreases wind and water erosion, (4) contain edge species with dense populations, and (5) maintain high species densities.
- **Corridors:** Natural vegetation corridors enhance important natural processes such as species and surface-water movement, while corridors of diverse fine-scale land uses result in efficient human and multihabitat species movement.

Incorporating these concepts into the physical landscape plan, however, requires a set of guidelines or a model planners can follow. As a direction for environmental planning and design, the "aggregate-with-outliers" model has been shown to possess several significant ecological benefits (Turner, 1989; Foeman, 1998). A range of benefits to human populations has also been suggested. These include:

- Providing a wide range of settings.
- Developing fine-scale areas where jobs, homes, and schools can be located in close proximity.
- Enhancing the efficiency of human movement along corridors.
- Introducing natural vegetation back into the development pattern.
- Providing for specialization within aggregated built-up areas.
- Enhancing visual diversity in the landscape.

Natural hazard considerations

Any review of the physical environment discussed within the context of planning would not be complete without a treatment of those physical processes and variables that represent hazards. While a more detailed discussion of hazard and risk assessment is presented later in Chapter 6, here, an outline of natural hazards and their basic characteristics is provided in order to: (1) identify critical environmental processes that produce hazards, (2) describe their salient properties and relationships, and (3) integrate environmental hazards into the planning process.

The concept of a natural hazard is somewhat misleading when removed from a human-centered view of the environment. Processes typically identified as natural hazards are nothing more than natural processes or events that have always been present in the landscape. These processes only become hazards when humans choose to occupy the same geographic location where such occurrences are common. Therefore, to a large degree the naturalness of these hazards becomes a philosophical and psychological barrier that can be difficult to reconcile when attempting to minimize their adverse effects. Further complicating the treatment of natural hazards is the observation that although natural hazards destroy property and often take human lives, they also perform vital environmental functions (Keller, 1988). For this reason understanding the nature of events that pose a hazard carries important implications that should influence environmental planning.

A list of natural events that can be assigned a hazard designation would include:

1 Geologic and geomorphic events:
  - Avalanches
  - Earthquakes
  - Erosion
  - Landslides
  - Tsunami
  - Volcanic eruptions
2 Climatic events:
  - Drought
  - Floods
  - Hurricanes
  - Tornadoes
  - Blizzards
  - Hailstorms
  - Severe thunderstorms
  - Fog
  - Heat waves
3 Biological processes:
  - Infestations
  - Fungal diseases
  - Bacterial diseases
  - Viral diseases

Detailed treatments of the nature and mechanisms that form these events can be found in the natural hazard literature (Murck, 1996; Blaike et al., 1994; Handmer, 1990). With specific reference to each of the processes listed above, several critical factors must be known if the environmental planner is to adequately address the nature of the hazard. Seven of the more essential descriptions of hazards include:

1 **Frequency** (interval and seasonality) – defines the time interval between events. Certain hazards have a definite seasonality associated with their occurrence, while others display temporal frequencies influenced by causal factors that operate from tens to thousands of years.
2 **Intensity** – describes the relative impact of the event as a function of its magnitude and local vulnerability.
3 **Duration** – explains the length of time the event is active. Some events are short-lived and may last for only a few seconds. At the opposite extreme are events such as drought that may least for several years.

4 **Spatial extent** – characterizes the size of the geographic area affected by the event.

5 **Exposure** – defines the population and infrastructure in proximity to the hazardous event that stands to affect it.

6 **Vulnerability** – explains the level of risk associated with the event given the pattern of exposure.

7 **Magnitude** – measures the physical power of the event as determined by a widely used metric such as the Richter scale for earthquakes, or the Fugitta scale for tornadoes.

Provided with this information, the planner has a clearer understanding of the physical processes that are active in the planning and which of those processes pose a hazard. Above all, this information serves as critical input to procedures used to evaluate both the present levels of exposure and vulnerability and future levels of risk associated with planned land-use changes. Information on the nature of hazards also greatly affects design considerations, building codes, densities, and the future allocation of sensitive land uses in zones that evidence hazards.

Physical systems and design

The preceding review of the natural factors that define the foundation knowledge that drives environmental planning serves to illustrate the importance of integrating earth-environmental science data into the process of making plans. The rationale supporting this form of integration is stated eloquently by McHarg (1969):

To learn of the evolution of the physical and biological processes is an indispensable step towards the knowledge one needs before making changes to the land; but it is far from enough. It is as necessary to know how the world "works." Who are the actors and how do they respond to the environment, to physical processes and to other creatures?

Ecological processes are therefore more than a planning philosophy, they provide the basis for planning and design if balance and harmony with nature are to be maintained. Central concepts such as the dependence of one life process on another; the interconnected development of living and physical processes of earth, climate, water, plants, and animals; the continuous transformation and recycling of living and nonliving materials, are more than the elements of a self-perpetuating biosphere, they are the central determinants of all human activities on the land (Hough, 1995).

Successful integration of the driving forces of urbanization with ecology is achieved through the design process. As suggested by Hough (1984), design is by definition a problem-solving activity and a mechanism that fosters integration. Design directs connections between disparate elements to reveal potentials that might not otherwise be apparent. To this end, several principles can be noted that, when brought together with the social and economic objectives which propel planning, effectively guide plan design and frame natural environmental factors as active decision criteria critical to the evaluation of development strategies (Hough, 1984).

Process

Processes are dynamic. The form of the landscape is the consequence of the forces that give rise to it; geological uplift, and erosion of mountains; the hydrologic cycle and the forces of water; plants, animals, and humans living the land. The form of the landscape, therefore, reveals its natural and human history, and the continuing cycles of natural processes that steadily influence its morphology. Understanding the dependence of form on process and recognizing that human and natural processes are constantly at work modifying the land illustrates the need to incorporate a process orientation into design. This prerequisite places design in the role of initiating purposeful and beneficial change, with ecology and people as its foundation.

Economy of means

The idea behind this principle has been stated in several different ways, yet each conveys the same general meaning. From the principle of least effort to the notion that small is beautiful, the economy-of-means principle encourages design to achieve the maximum output from the minimum input. This suggests that simplicity, while a relative term, and balance with natural process should be common to any planning solution.

Diversity

In a purely ecological context, diversity implies richness and is used to express either the number of species in a given area or the genetic variability within a single species. If health can be described as the ability to withstand stress, then diversity may also be used to imply health (Hough, 1989). Diversity also suggests choice and a mixture of features from which selections can be made, be those land uses, employment opportunities, lifestyles, or landscapes. In design terms, diversity translates as the creation or preservation of a mixture of settings that can enhance interest and aesthetic appreciation, and stimulate social health.

Environmental connectedness

In an ever-urbanizing world there is a need to maintain a sense of connectedness with the environment. The natural environment provides one of the most appropriate, flexible, and diverse settings for connecting people to place. The constant and direct experience of how the environment works, assimilated through the daily exposure to and interaction with the landscape, should be a priority of design. There is nature in the city that can be amplified through design, and a need exists to give emphasis to the natural systems that operate in the urban landscape.

Change as constructive process

The natural tendency in planning, particularly with respect to environmental planning and design, is to focus attention exclusively on designs that minimize destruction of environmental resources. Although the minimal-impact approach is the obvious response to human-induced change, it suggests an acceptance of negative values (Hough, 1984). This position can inhibit more creative solutions that develop from ecology-based design. Thus, design needs to approach the question of how human development processes can contribute to the environments they change. Change becomes a positive force capable of enhancing the landscape through habitat-building, recycling wastewater, storm water conservation, and other related methods. In this way development manages and reuses the resources it draws on. Development becomes more than simply something superimposed onto the landscape, it becomes linked to existing natural processes.

The natural factors discussed in this chapter and the design principles that grow out of them draw the environment firmly into the planning process. However, knowledge of the environment has to be supported by the acquisition of site-specific environmental information and methodologies that can evaluate that information in relation to the problems confronting the planning area. Landscape characterization and assessment applies what we have learned in this chapter and connects us to alternatives that guide the future use of the landscape. In the next chapter the issues and methods that direct land evaluation and the characterization of the landscape are examined.

Summary

To successfully integrate environmental concerns into the environmental planning process, the planner must possess a basic understanding of the environmental processes active in the planning area. With this information, the planner can better comprehend the importance of critical environmental features and discern the influence they have on qualities of the built environment. This chapter presented a review of the fundamental "what I need to know" information pertaining to the salient natural factors that direct environmental planning. Beginning with the geologic environment and culminating with an overview of natural hazard considerations, the role of each factor in landscape design and the dynamics of the processes involved were identified. Because development is more than simply the superimposition of built form onto the landscape, how development becomes linked to existing natural processes becomes a central question to resolve in order to maintain environmental sustainability.

Focusing questions

Why assess the physical environment of the planning area; what significance does this assessment convey?

In what ways do the natural factors reviewed in this chapter exert controlling influences on development?

How does one plan within the context of a dynamic physical environment?

In what ways does environmental process suggest options to direct and form the design of development proposals?

CHAPTER 5

Landscape Inventory and Analysis

Information is a critical component of any decision-making process, and environmental planning is no exception to this rule. As demonstrated by Lein (1997), every decision-making body relies on information to exist, and the acquisition, processing, and dissemination of information are fundamental activities undertaken by planning agencies. Information can be considered a form of planning intelligence, and planning intelligence is a type of strategic decision support information that enables the environmental planner and the community to identify, understand, and manage the problems characterizing the planning area. This intelligence is derived from organized data and facts about the regional setting and those features that are of specific concern to planning. In this chapter the methods used to collect and organize this information, and the evaluation procedures employed to provide the "intelligence" needed to direct the environmental planning process, are examined.

Regional landscape inventory and monitoring

Inventory and monitoring are among the most basic tools that enable environmental planners to establish critical baseline information and measure change. Unfortunately, information pertaining to the natural environment has often been used in an *ad hoc* manner (Steiner, 1991). There is a tendency to collect only that information required to address a specific problem. That information, because it is specialized, is often disconnected from the larger concerns of the planning area and fails to satisfy more generic information needs. This results in a one-dimensional view of the landscape that does not support the ecological premise that everything is connected to everything else (Steiner, 1991). Yet, without providing a more systematic view of the landscape, "seeing" the connections between the natural elements of the landscape and gaining insight regarding their interaction is difficult to achieve. Furthermore, because environmental planning is based on an integrated view of the landscape, synoptic-scale data concerning the biophysical processes characterizing the region must be collected and analyzed to provide the baseline knowledge from which plans can be developed (Steiner, 1991). With these elements in place, the data gathered through inventory and monitoring activities helps to substantiate environmental planning and improves the credibility of the environmental plan. Deriving the information needed to reach this point, however, depends on the implementation and timing of programs developed to inventory and monitor the regional landscape.

An inventory, as its name suggests, is an itemized listing of current assets. If we consider the geographic area of a watershed, it would be necessary to have information pertaining to that area's geology, climate, land use, land cover, and vegetation, as well as a range of other factors, so that effect plans to manage the watershed can be drafted. Thus, whether in the context of environmental planning or with reference to the local retail department store, the inventory represents a basic accounting of the presence or status of important and valuable resources whose present disposition is unknown. Through the inventory process an understanding of these assets is acquired and their quality, quantity, condition, and location become known. As a preplanning mechanism, the inventory plays a vital role by providing the detailed overview of the region necessary before specific programs or policies can be drafted. In this role, the inventory describes the salient environmental variables that characterize the planning area, identifies their geographic extent, and carefully documents their important properties. Once completed, the inventory forms a baseline information source against which plans and programs can be evaluated, and provides a mechanism whereby the opportunities and constraints of the environment can be disclosed and recommendations can be offered to guide future development within the region.

For an inventory to be effective several organization questions need to be addressed. Taking each question in turn, they help to refine the methods used to develop the inventory and ensure its adequacy.

What to inventory?

Unlike our department store analogy with items arranged on stockroom shelves, the natural resource inventory is complicated by the multidimensional nature of the environment. The relevant factors to include in an inventory may be predetermined by local, state, or federal mandates. In other instances selection of the natural factors to inventory may require careful deliberation by the planner. Selecting the appropriate factors to inventory depends largely on purpose. In the majority of cases an inventory is conducted to give decision-makers a comprehensive and detailed understanding of the physical and human characteristics of the planning area. Using this information-base, environmental data can be

employed directly in setting development options and priorities, identifying problems, and increasing general knowledge about the region that can be used by decision-makers.

Typically the dominating or prominent features of the landscape are chosen and described in a manner such that a decision-maker unfamiliar with the planning area can learn enough about its important qualities to ensure that plan development and implementation remains focused. Natural Factors given consideration in a resource inventory may include the follow items:

1 Habitats found within or adjacent to the planning area.
2 Major cultural resources found within or adjacent to the planning area.
3 Major land uses and related land-use activities found within or adjacent to the planning area.
4 Major land and resource ownership patterns and management responsibilities.
5 Major historic, prehistoric, and archaeological resources.

A comprehensive outline of the natural, cultural, and amenity resources that can typically be examined and subject to inventory are presented in Table 5.1. Describing each of the factors presented in the table may involve the production of statistical summaries, detailed maps documenting location and geographic arrangement, narrative summaries reviewing and explaining important features or relationships, photographs and other supporting graphic displays to enhance visualization of the planning area. Because the overriding goal of the inventory is to present data and information to improve understanding, anything not related to improving communication and the transfer of critical environmental knowledge should be avoided.

How to inventory

An inventory can be derived from a mix of primary, secondary, and tertiary sources. Consequently, data collection and data evaluation become critical phases in the inventory process. Perhaps the most important step in data collection involves identifying the boundaries that will delineate the

Table 5.1 Landscape factors subject to inventory.

Natural elements
Physiography: slope, drainage, unique features
Geology: bedrock formations, faults, fractures, slumps, slips
Soils: type, composition, permeability, erosion potential
Hydrology: drainage systems, springs, seeps, wetlands
Vegetation: plant associations, unique communities
Wildlife: habitats
Climate: temperature, precipitation, wind flow, humidity,
   evaporation

Cultural elements
Transportation: roads, rail lines, airports
Utilities: oil, gas, water networks
Structures & excavations: buildings, mines, dumps
Ownership: name, assessed value
Historical/archaeological: significant features and sites

Visual and aesthetic elements
Major viewsheds
Minor viewsheds
Scenic areas
Unique features
Points of interest

planning area and delimit the geographic extent of the biophysical and sociocultural factors that will be mapped and described. Although boundaries may be set by legislative goals, there is a hierarchical quality to an inventory that directs description and mapping through different levels of geographic scale and resolution. Generally, an inventory proceeds from a regional scale to a local level of detail. This hierarchical structure is used to place the planning area into its larger environmental context and to enhance the decision-maker's ability to see relationships that may exist between the planning area and the larger regional landscape. The drainage basin at the regional level and the watershed (or sub-basin) at the local level serves as the ideal natural delimiting feature for ecological planning and analysis (Steiner, 1991; McHarg, 1969). Recall that a drainage basin is identified by locating the drainage divides between channels of a particular order through the use of contour (elevation) data and runoff patterns. Since the drainage basin is delineated on the basis of physiographic and hydrologic criteria, it creates a practical reference for connecting natural and cultural attributes that facilitates system representations. The drainage perimeter has also

been used to approximate ecosystem boundaries. By applying the drainage basin as the primary organizing spatial unit, data characterizing the natural, cultural, and amenity resources presented in Table 5.1 can be collected and mapped.

Cartographic treatment of inventory data is an important method of communication, visualization, and analysis by which the quality of the inventory can be determined in large part by the quality and accuracy of the maps created for display. Therefore, one of the more critical decisions that must be made early in the inventory process has to do with the selection of a base map and scale of representation. The base map will be used to anchor all the spatial information collected for the inventory. Factors that will influence the appropriateness of the base map include: (1) scale, (2) projection, (3) locational reference, and (4) medium.

1 **Map scale** – a map is a scaled representation of a portion of the earth's surface. The scale of a map is simply the ratio between map distance and earth distance. However, scale greatly influences the level of generalization required in order to represent geographic objects and directly controls the degree of detail that can be seen on a map. It is therefore essential that an appropriate scale is selected to that the representation of the surface can be as accurate as possible and the features and patterns of natural factors shown without unnecessary abstraction or generalization. Although the increased use of digital geographic information systems (GIS) has reduced some of the concerns regarding scale, every map product, whether stored in a machine environment or printed on paper, will posses a level of resolution (the smallest ground area that will be treated as an object that can be represented on the map) as a function of its scale that will direct how objects can be represented and symbolized.

2 **Projection** – the characteristics of a particular map are determined by the projection on which it is plotted. The retention of shape, equivalence, direction, or distance are important considerations when deciding upon the purpose of a map. Retaining one charac-

teristic generally results in the distortion of other potentially important characteristics. For mid-latitude regions, conic projections such as the Lambert or Alber's projection are commonly used on base maps, although other options are possible.

3 **Locational reference** – Before maps can be prepared that actually portray the locations of surface objects, a means of determining location must be selected. Locational systems can include latitude and longitude as well as more specific reference grids such as the Universal Transverse Mercator (UTM) grid system, or the State Plane Coordinate System.

4 **Medium** – traditionally mapped data were compiled and printed on paper or polyester film. While maps placed on these media are still very common and useful, the emergence of digital cartography and GIS together with the advent of the worldwide web (www), has introduced new source material for map compilation and display. Selection of map media must be taken into consideration since this simple decision will influence the ease of data storage, stability, accessibility, permanence, and ease of update and error correction. While inventory data may be collected from a variety of sources including field surveys and sampling, data are likely to be assembled and compiled into a digital form for computer storage and processing. For this reason, digital formats are preferred and provide greater flexibility when compared to traditional map media.

With a base map selected, the inventory can proceed as an exercise in data collection, data compilation, and geographic description and mapping. Since the inventory explains the first technical phase in the sequence on which planning depends, the compilation of inventory elements is directed by the information content of each data item. To be useful, it is essential that the information presented in the inventory meets three fundamental requirements (Gonzalez-Alonso, 1995). First, it must be comprehensive and gaps in the information should be avoided wherever possible. Secondly, the inventory must

be systematic so that it can be applied easily. Lastly, the inventory must be multidimensional and give reasonable consideration to the totality of the significant environment.

The key decisions affecting how an inventory may be compiled relate to the choice of elements and the appropriate level of detail that will effectively communicate their importance. The important factors to consider when planning the inventory include: (1) the general characteristics of the planning area, (2) the objectives and socioeconomic implications of the inventory, (3) the size of the planning area, and (4) the availability and quality of data. With reference to information content, processing, and purpose, a resource inventory can be targeted at three interrelated levels:

- A reconnaissance level appropriate for describing regional patterns or elements that characterize large-scale regional trends.
- A semi-detailed meso-scale level oriented toward general planning questions with more specific data needs.
- A detailed site-specific level required for localized analysis involving siting decisions and the assessment of environmental impact.

Although exhaustive collection of information may not be required in all cases, concentrating on those elements that supply information that define the planning area makes it possible to map elements into homogeneous regions, and define elements in a clear and simple manner (Gonzalez-Alonso, 1995).

Documenting the inventory

The data selected and collected in the inventory must be assembled into a format that will give decision-makers a clear indication of the landscape's significance and outline the main implications that should direct future use. Therefore, the inventory and the recommendations that grow from it provide a basis for subsequent analysis and for raising questions concerning specific features of the landscape, such as:

- its susceptibility to modification
- its stability
- its aesthetic qualities

- its conservation value
- its opportunities and constraints.

Based on these points, it can be seen that an inventory represents more than simply an exercise in data collection. It provides an initial assessment of site characteristics that allows the potential of an area to be critically examined. Therefore, to be useful, a detailed evaluation of each factor together with a series of planning recommendations should accompany the discussion. This form of cursory evaluation identifies the issues, concerns, possibilities, and limitations that are associated with each natural factor, and allows options to be entertained without specific goals or objectives in mind that might otherwise limit thinking.

With respect to documentation, the ultimate goal of an inventory is to produce an organized and concise review of the pertinent material that clearly describes the planning area in both a narrative and graphic form. An inventory is typically divided into five major sections beginning with an introduction and proceeding systematically through each thematic factor. A generic outline presenting the basic structure and contents of a natural resource inventory is given in Table 5.2. Each section in this document has a specific purpose and establishes the background needed to proceed to the sections that follow. The logic behind this design is simple. Beginning with the introduction, an overview of the planning area is offered and the purpose and scope of the inventory is explained. Information presented in the introduction establishes the site and situational context of the planning area and helps communicate a sense of place. With the general characteristics of the regional setting described, the decision-maker has a reconnaissance-level perspective of the planning area. Moving from this general overview, the section that follows can introduce the human and cultural factors that define the area, along with a brief historical treatment of the region to document its origins and suggest its evolution. In this section emphasis is also given to aesthetic qualities and known management problems that exist within the planning area. The problems identified in this section can include a range of social, economic, and environmental issues that should be recognized as the information in the

Table 5.2 General resource inventory contents.

I. Introduction
   a) site description
   b) purpose of inventory
   c) table of contents

II. History of Site and General Information
   a) land ownership patterns
   b) existing land use
   c) aesthetic qualities and special features
   d) general planning recommendations

III. Natural Site Conditions
   a) geology
   b) vegetation
   c) soils
   d) hydrology
   e) climate
   f) wildlife habitat

IV. Summary of Planning Recommendations
   a) review of planning considerations
   b) recommended uses
   c) development constraints
   d) other limitations

inventory is interpreted. Using this discussion as a backdrop, this section of the inventory may conclude with a review of general planning recommendations that have been derived from the analysis of data inventoried. Although a more detailed treatment of these recommendations will follow later in the document, a general summary is presented here to help frame the material that is introduced in the sections to follow.

The next major section of the inventory can be considered to form the heart of the document. In this section the detailed information pertaining to the natural site conditions selected for examination are presented. While the sequence or order may vary, generally a "ground-up" approach is used. This approach allows the environmental factors that define the planning area to unfold in a manner that facilitates an understanding of factor interactions and interrelationships.

The final substantive section of the inventory contains recommendations regarding the present state and the potential concerns surrounding each factor. This section is critical to the success of the inventory, particularly if it is to have value as a basis for goal-setting and plan-making. Recommendations address the particular types of use, pattern of development, design concerns, and locational constraints that where identified during the inventory and analysis process. The concept of a constraint is particularly important to this purpose. Constraints can explain environmental limitations evident with respect to a given factor, from expansive soils, flood risk, slope instability, and sensitive habitats, to historic sites, cultural amenities, and areas of scenic value. Each constraint is documented and the overall developmental suitability of the planning area for future use is described in relation to how each factor imparts its influence. Although recommendations are merely suggestions, they often identify important alternatives that can be better accommodated by the site given its biophysical and socioeconomic characteristics. Moving the inventory into a more coherent expression of opportunity or constraint directs our attention to the methods of land evaluation and analysis.

Land evaluation and analysis

A large tract of land is being considered for some form of use. The question is what use is best suited for that area. This is the general question that guides the land evaluation process and the methods that have been developed to assess the "fitness" of land for development. However, numerous environmental factors influence the development process and require careful evaluation. From topography, surface drainage, climate, and soil, to the incidence of natural hazard, and disease, a range of environmental characteristics affect the current as well as the future state of the landscape system and impart some direct influence on the land use placed at a given location. Consequently, there are considerable benefits to be realized from a simultaneous assessment of the landscape factors critical to the sustainable use of the planning area. The regional landscape inventory is the first phase in this more comprehensive process of landscape evaluation.

Drawing from the fundamental principles that (1) land should be used for the purposes for which it is best suited and that (2) land of high value for

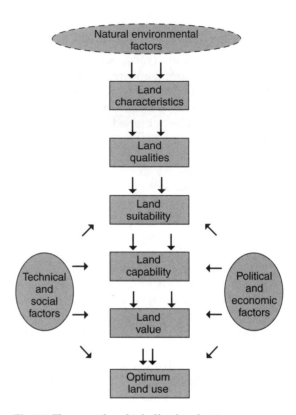

Fig. 5.1 The general method of land evaluation.

an existing use should be protected against changes that are difficult to reserve, the natural environmental factors that define the regional landscape are subjected to six interpretative stages that when completed yield an expression of the optimum configuration of land use within the planning area (McRae & Burnham, 1981). The stages defining this evaluative process are illustrated in Fig. 5.1. By definition land evaluation is the process of estimating the potential of land for alternative uses (Dent & Young, 1981). Through this evaluation process a comparison is made between the requirements of specific types of use and the resources offered by the environment to support them. Fundamental to the evaluation procedure is the widely understood fact that differing land uses will define differing requirements. Since nearly all human activities use land to some degree, the opportunities and limitations of this increasingly scarce resource must be translated into a form that can aid decision-making (McRae &

Burnham, 1981). Land evaluation is the methodology designed to achieve this objective.

As a general methodology, land evaluation can be conducted directly by trial and error, or indirectly through the application of an analytical approach that attempts to explain or predict land performance under specific categories of use. The logic that makes land evaluation possible relates to the fact that land varies in its physical and human-geographic properties. Therefore, this inherent variation should affect land use, suggesting that given a proposed use, some land areas are more appropriate in physical and/or economic terms. This observable variation is in part systematic with definite and knowable causes. Because causal influences can be known, the variation, whether expressed in physical, political, economic, or social terms, can be mapped. Furthermore, the behavior of the land when subjected to a given use can be predicted with some degree of certainty, depending on the quality of data and the depth of knowledge relating land to land use. Hence land suitability for various actual and proposed uses can be systematically described so that decision-makers can apply these predictions to guide their decisions.

Providing this specialized information is the function of land evaluation and information becomes a key ingredient in this process. Typically three sources of information are required: (1) land, (2) land use, and (3) economics. Land data is obtained directly from the natural resource inventory. Information regarding the ecological and technical requirements of various land-use types is acquired from the subject areas that have developed applied and theoretical knowledge related to each category. Economic data may not be necessary if the results of an evaluation are needed to address purely physical issues; then only general economic and social patterns are required. However, if the results are required in strict economic terms, then data on specific costs and prices is needed (Dent & Young, 1981). These data needs coupled with the observation that direct evaluation methods tend to be of limited value unless large amounts of data can be collected, support the use of indirect evaluation techniques (McRae & Burnham, 1981).

As an analytic tool, a land evaluation may be presented in terms that are qualitative, quantitative, physical, or economic (Dent & Young, 1981). A qualitative evaluation is one in which the suitability of land for alternative purposes is expressed in qualitative terms only. For example, observation of land areas based on soil, slope, and distance criteria may suggest that a given land area, because of its soil type, steepness of slope, and distance from major roads, may be of marginal value for a particular type of crop production. Therefore, the use of terms such as highly, moderate, or marginally suitable or not suitable when considering a specific type of use imply qualitative judgments have been made to place land into one of these categories. Typically, qualitative evaluations are employed mainly at a reconnaissance scale, or as a preliminary to more detailed investigations. While the results of this type of land evaluation may be highly generalized, it permits the integration of many environmental, social, and economic factors.

Quantitative physical evaluations provide numerical estimates of the benefits to be expected from the use of land. Quantitative evaluation is most frequently carried out as the basis for economic evaluation, where results are expressed in terms of profit and loss. Here, monetary values are applied to data from quantitative measures to obtain cost expressions and to derive profit and loss treatments of inputs and outputs. However, Dent and Young (1981) note that an economic evaluation is not strictly confined to profit and loss considerations. Other consequences such as the environmental or social can be included and combined with economic data as a basis for decision-making.

Regardless of the type of evaluation conducted, the goal of evaluation is to predict change and to provide better insight concerning the consequences of a development plan. Thus, as a planning tool, land evaluation becomes a necessity wherever changing the landscape is contemplated (Dent & Young, 1981). Prediction, in this context, directs attention to the concept of suitability. Generally, all land evaluation procedures attempt to express this idea and relate it either qualitatively or quantitatively to alternative types of land uses. Therefore, given different forms of use, with varying environmental opportunities and constraints, land evaluation methods strive to produce an expression of suitability that can be used to direct how land should be utilized. Using this expression, a comparison is made between the requirements of the contemplated land use(s) and the qualities of the land that are present to support it. The suitability of land can be assessed and expressed in several ways. Three of the more common include: (1) capability, (2) suitability, and (3) value. A variety of analytic techniques have been introduced to perform regional landscape analysis and land evaluation that apply these terms as a control which directs the future use of land resources (McAllister, 1986; Anderson, 1980; Steiner et al., 1994; Davidson, 1992).

Methods of landscape assessment

Techniques for regional landscape evaluation and analysis are general tools that assist with the task of assembling a large number of important landscape variables into an expression that can be used to direct decision-making. Through integration, data from diverse sources can be interpreted, evaluated, and communicated in a manner that guides the process of choosing among alternatives and identifying optimal patterns that maximize benefits and reduce or avoid costs. More importantly, these techniques complement the method of regional landscape analysis applied in environmental planning. This method, derived from the procedures outlined previously by McHarg (1969) and Steinitz (1977), define a combination of procedures which form a model that describes the general process. The four main phases of this model are as follows:

1 *Inventory* – describing the compilation of data on natural environmental conditions to serve as critical baseline information on which further analysis may draw.
2 *Data interpretation* – explaining the process of categorizing inventory data to define patterns and trends.
3 *Data synthesis* – defines the combination of

data through the use of specific analytic techniques and models to derive expressions of suitability or constraint.
4 *Plan formulation* – explaining the integration of land evaluation and ecological assessment results into the policy-making process.

Connecting data to the plan is accomplished through the application of one or more evaluative (analytic) techniques. With each technique adopted, certain assumptions or conventions are assumed that influence how well a given method can be applied and its results interpreted (McAllister, 1986). When one is conducting an assessment of a plan or design, assessment typically implies that a comparison is being made between two states, one of which is the condition that will result from implementing the plan. The effect of the proposal can be examined by comparing:

1 The original state existing before the action was taken, referred to as the baseline state.
2 The state that would have evolved in the absence of the action.
3 A goal or target state.
4 The ideal state.

The appropriateness or suitability of a proposal, therefore, relates to how well the various states compare. Thus, given baseline conditions, one asks how closely the goal state, ideal state, and the "non-action" state agree.

Not all assessment methods provide the same level of comparison. Generally, differences among methods are revealed by the way they address concerns related to the categorical descriptions used to estimate and report natural factors and effects, the manner by which magnitude and significance of change is estimated, and the procedures used to derive ratings intended to indicate the relative importance of each factor. With reference to the description of factors included in the selected methodology, the fundamental rule for selecting categories is that they must be mutually exclusive and exhaustive. Concerning the question of magnitude and significance, mathematical or statistical procedures are considered to be the most reliable means of estimation. However, in both cases, assessment relies heavily on the use of expert judgment. This is particularly the case in situations where: (1) scientifically valid procedures are lacking, (2) data needed to implement a procedure does not exist, or (3) where cost factors prohibit the use of certain procedures. The question of measurement can be addressed in a number of ways. First, the notion of rating implies either a subjective or quantitative ordering of a variable. Therefore, one distinction considers the source of the rating. Ratings can be derived from a variety of sources: measurable physical characteristics, monetary values, or expert judgment (McAllister, 1986). Finally, rating type describes the general approach that was used to scale qualities, quantities, or effects. Four scaling methods frequently applied in environmental assessment include (1) simple, (2) constant, (3) scaled value weight, and (4) rescaled weight. The general characteristics of these rating schemes are summarized in Table 5.3. A more detailed discussion of these approaches can be found the McAllister (1986).

Recognizing the distinctions that exist among various assessment techniques available helps the planner identify the approach that will produce the most meaningful results. In general an assessment technique should be:

- **Systematic** – any technique must be systematic to ensure that the results it provides are replicable.
- **Simple** – the techniques selected should not be overly complex.
- **Quick** – the techniques must be able to generate useful results within a reasonable length of time.
- **Inexpensive** – the techniques should not impose unnecessary costs.
- **Legally acceptable** – the techniques should conform to the various legal and administrative requirements to which it is subject.
- **Comprehensive** – the techniques should be capable of incorporating all of the relevant factors important to the decision-maker.

While strict adherence to these points may not be possible in all situations, they focus attention on the practical aspects of the analysis problem and offer some insight to help evaluate whether a certain technique is feasible or not. The methods available to the environmental planner fall within four broad categories: (1) methods of land capability analysis, (2) methods of developmental suitability analysis, (3) methods of carrying capacity analysis, and (4) methods of land evaluation and site assessment. Each of these techniques is reviewed below.

Table 5.3 Characteristics of land rating schemes.

Scaling method	Characteristics
Simple rating method	A set of guidelines or standardized procedures are followed in assigning judgmental importance scores
Constant value weight	Rating applied to each unit of a given type of impact
Scaled value weight	Derived from a mathematical function, they are used to avoid possible inaccuracies of constant weights
Rescaled impact weight	Scaled weights adjusted by expert judgment; used to overcome factor dependencies problems

Land capability analysis

Land capability analysis is a technique designed to classify land units based on their ability to support general categories of use. The term capability has specific meaning in this context. When used in reference to land, capability implies that a given area, due to its inherent characteristics, may be qualified for a specific type of use (i.e. agriculture, open space, urban). Determining the qualifications land must possess to be considered appropriate relies on relating the land area to a set of environmental factors that influence how well a given land use will perform. The environmental factors typically employed in capability analysis include soil erosion potential, susceptibility to flooding, climate, and slope stability, together with other factors that stand to degrade the utility and productivity of land. The analysis of land capability, therefore, requires an evaluation of the degree of limitation posed by permanent or semi-permanent attributes of land to one or more land uses (Davidson, 1992). The logic guiding analysis is comparatively simple. Capability is based on the premise that as the degree of constraint increases, land should be allocated to lower categories or less intensive forms of use. Although this technique was originally developed to assist with agricultural planning and designed to help identify agricultural land uses that would not contribute to environmental degradation, the concept of capability is flexible enough to permit broader interpretations.

There are two principal ways whereby land capability can be assessed. One example is through the use of categorical systems. Categorical systems are implemented by testing the values of appropriate soil and site properties against criteria for a set of land-use categories. Site conditions are compared against these evaluative criteria and if the minimum criteria are not met, then the land area in question automatically falls to the next class and the process repeats until a match is identified. Land characteristics are then tested against less stringent criteria as the evaluation process continues through the classification system until a category is identified where all the criteria are satisfied. A second means of deriving an estimate of land capability is through the use of parametric systems. Parametric systems apply mathematical formulas to combine site properties into a quantitative expression. Although parametric systems differ in respect to the factors that are included for mathematical manipulation, three general approaches are common:

1 Additive systems of the form $P = A + B + C$
2 Multiplicative systems of the form $P = A \times B \times C$
3 Complex systems of the form $P = A (B + C) \times D$.

In each of the examples listed above, P is the parametric rating or score and A, B, C, and D represent soil and site properties. The basic features of categorical and parametric approaches are examined below.

Capability classification

A variety of land capability classification techniques have been introduced (Davidson, 1992). Perhaps the best known of these is the USDA method formalized by Klingebiel and Mont-gomery (1961). The USDA method employs a three-tier structure in its classification system:

- **Tier 1, Capability Class** – the top level in the classification with a total of eight classes defined and labeled I to VIII inclusive, indicating the degree of limitation in descending order of severity.
- **Tier 2, Capability Subclass** – indicating the type of limitation encountered within the class, these subclass designations identify limitations such as erosion hazard, climate, fertility, or excess water that restrict the use of land. The limitations recognized at the subclass level are:

  Subclass e (erosion), defining soils where susceptibility to erosion is the dominant limitation.

  Subclass w (excess water), explaining conditions of poor soil drainage, wetness, high water table, and overflow as the dominant limitation.

  Subclass s (soil), describing soils that have limitations associated with shallowness, low moisture-holding capacity, or salinity.

  Subclass c (climate), identifying land where climatic conditions limit land use.
- **Tier 3, Capability Unit** – a subdivision of the subclass where the variation in degree or type of limitation is similar across soil type. Thus within the same narrow range of environmental conditions, capability units define land areas that share a similar response to management and improvement practices.

Soil and climate limitations in relation to the use and management of soils are the basis for differentiating capability classes. Assigning land areas to capability units, capability subclasses, and capability classes is conducted on an evaluation of 8 environmental factors. These are summarized in Table 5.4. The assignment process itself is based on a series of assumptions that have been discussed in detail by Klingebiel and Montgomery (1961). When carefully applied to a well-defined problem, land areas falling within one of the eight classes can be described and mapped accordingly (Davidson, 1992):

- Class I: soils with few limitations that restrict their use.

Table 5.4 Criteria used for placing soils into capability classes.

1. Ability of the soil to give plant response to use and management based on organic matter content, ease of maintaining nutrients, percentage base saturation, cation-exchange capacity, parent material, water holding capacity
2. Texture and structure of the soil to the depth that influences the environment of roots and the movement of air and water
3. Susceptibility to erosion
4. Soil permeability
5. Depth of soil material to layers inhibiting root penetration
6. Alkalinity
7. Physical obstacles
8. Climate

- Class II: soils with some limitations that reduce capability.
- Class III: soils with some severe limitations that require special management.
- Class IV: soils with very severe limitations that restrict certain activities and require special management.
- Class V: soils with little or no erosion hazard but with other limitations that restrict extensive activities.
- Class VI: soils with very severe limitations and that are unsuitable for cultivation and restrict use to pasture and range.
- Class VII: soils with very severe limitations that make them unsuited to cultivation and restrict use to woodland or wildlife.
- Class VIII: soils and landforms with limitations that preclude their commercial viability and restrict use to recreation, aesthetic, or watershed purposes.

Implementing the USDA method is largely subjective since the criteria for establishing class limits are not generally specified. Therefore, the technique can be considered a formal representation of expert-technical judgment. Although the results can be useful, the advantages and disadvantages associated with land capability classification must be well understood in order to guide its appropriate application. The strengths and limitations of the USDA method have been examined by McRae and Burnham (1981) and are summarized in Table 5.5. Problems related to

Table 5.5 Advantages and disadvantages of the USDA system.

Advantages:
- Division into small number of ranked categories is easily understood
- Qualitative and suggests realistic approach given knowledge limitations
- Versatility
- Easily applied
- A general-purpose classification system
- Stresses adverse effects of poor management
- Useful way of relating environmental information
- Results easily displayed on maps

Disadvantages:
- Subjective
- Interactions between limiting factors difficult to express
- Division into too few categories overgeneralizes conditions
- Implied ranking suggests true land value
- Fails to include socioeconomic factors
- Emphasizes limitations rather than land's positive potential
- Difficult to use where reliable soil information is lacking

imprecision and subjectivity must be balanced, however, by the flexibility such categorical systems offer. Flexibility of use has contributed to the design of similar measurement frameworks in Canada, Great Britain, and the Netherlands, each modeled after the USDA approach (Davidson, 1992).

Parametric methods

Dividing land into a small number of categories that are mutually exclusive and supposedly exhaustive imposes an artificial structure that does not correspond well with the variability inherent to complex environmental interactions. By relaxing the need for discrete classification and replacing nominal categorization of land with continuous scale assessment, more realistic results can be obtained (McRae & Burnham, 1981). Applying parametric techniques in a land capability analysis involves:

1. Establishing a land unit to assess.
2. Obtaining the required data.
3. Developing the appropriate measurements for that data.
4. Translating the raw measured data into a coding or rating scale.

Table 5.6 Calculation of the Storie Index Rating System.

Index parameters:

Factor A – Rating based on characteristics of physical profile
Factor B – Rating based on surface texture
Factor C – Rating based on slope
Factor X – Rating based on other site conditions

Index calibration:	Percent:
Soil type X – brown upland soil with shale parent material and depth to bedrock at 90 cm	Factor A = 70.0
clay loam texture	Factor B = 85.0
rolling topography	Factor C = 90.0
moderate sheet erosion	Factor X = 70.0

Index rating:
SIR = 0.70 × 0.85 × 0.90 × 0.70 = 0.37 or 37%

Source: Adapted from McRae and Burnham (1981).

5 Performing the desired mathematical combination of measured properties.
6 Applying the resulting scores to the entire planning area.

A selection of additive, multiplicative and complex schemes has been reviewed in detail by McRae and Burnham (1981). Perhaps the most widely known parametric method is the Storie Index Rating System (Storie, 1978). This multiplicative system computes a land-quality rating index based on the general relation

$$SIR = A \times B \times C \times D \qquad (5.1)$$

Where:
A = character of the soil profile
B = texture of surface soil
C = slope
D = miscellaneous factors (i.e. drainage, alkalinity).

Ratings for each factor are provided and expressed as percentage scores in Storie (1978). An example of the scores applied in a hypothetical calculation are shown in Table 5.6.

As with the categorical approach, parametric systems possess important advantages and disadvantages that should be understood before these methods are applied. The main advantages of these systems when compared to categorical methods include:

- reduced subjectivity
- ease of adaptation
- attractive simplicity
- amenability to computer manipulation.

The main limitations are related to concerns surrounding

- misleading impressions of accuracy
- calibration difficulties
- parameter inconsistencies.

An alternative means of evaluating land shifts the focus from capability to the concept of developmental suitability.

Developmental suitability analysis

Developmental suitability may be defined as the "fitness" of a given tract of land for a well-defined use (Steiner, 1983). When compared to the concept of capability, suitability explains the condition where a singular use is described that is the most appropriate use for a site in the landscape. Therefore, while capability narrows the search for alternative land uses that may be "qualified," suitability refines this search further to identify the single use that fits with the environmental conditions present at a given location. To illustrate this important terminological distinction, consider the example of a county that is exploring options for currently undeveloped land areas. The county wishes to identify what uses might be appropriate and seeks workable alternatives. Through the application of land capability analysis, broad categories of general land use are identified that suggest where land may be qualified to support urban, agricultural, and recreational uses. Within each of these categories, developmental suitability analysis is performed to determine which specific types of urban, recreational, and agricultural uses are the most appropriate. Thus within the general land-use category of urban, suitability analysis can suggest which areas may best support residential, commercial, or industrial forms of development.

Variations in the degree of suitability are determined by the relationship, actual or anticipated, between benefits and the required inputs associ-

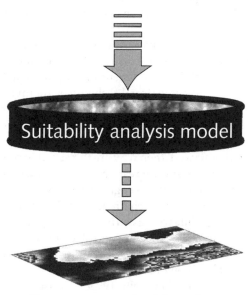

Landscape attributes

Suitability analysis model

Recommended land uses

Fig. 5.2 The suitability analysis "filter."

ated with the use in question for the tract of land under consideration (Brinkman & Smyth, 1973). Focused on the concept of suitability, effective analysis depends on the techniques that make systematic use of environmental information and relate the condition of the environment to the activity it is proposed to accommodate (Ortolano, 1986). In this context, suitability analysis can be looked upon as a filter, where land-use requirements are passed through an environmental screen. Land uses passing through this screen can be considered the most suitable for the site under consideration (Fig. 5.2).

Identifying the potential opportunities or constraints of a site requires selection of those landscape attributes that would support or influence the functioning and sustainability of the activity under consideration. A variety of techniques have been proposed to perform a suitability assessment (Hopkins, 1977; Fabos, and Caswell, 1977). Traditionally, selected landscape attributes are depict-

ed in map form and combined cartographically so that when viewed collectively they provide insight into the type of use intrinsically suited to that location. This generalized procedure for conducting a developmental suitability analysis has been described by Lein (1990). In nearly all cases, obtaining an expression of suitability becomes a function of the analyst's associative logic when relating landscape qualities as indicators of constraint to a specific developmental design. As noted by Lein (1990), this method of analysis contributes to the common practice of compiling a series of data maps each representing a selected environmental theme or variable, interpreting a combined pattern, and drawing conclusions from their composite profile. However, the practical limitations of complex manual overlaying operations, coupled with the inability to consistently confirm the results obtained through visual inspection of superimposed map surfaces, gave way to alternative procedures (Hopkins, 1977; Steiner, 1983; Bailey, 1988).

At present, 7 general approaches to developmental suitability analysis have been introduced (Table 5.7). With some variations, an algorithm is typically applied to a set of environmental variables and produces an expression of suitability through either direct or implied mathematical operations. The environmental variables selected for inclusion in the suitability algorithm are assigned some type of value, rating, score, or weight. Using these numerical approximations of fitness, developmental suitability is derived as a function of cross-factor addition or multiplication of these numerical expressions as dictated by the respective algorithm. The cross-factor results are then assembled into an arbitrarily derived classification scheme to allow a portrait of developmental suitability to take form (Lein, 1990).

Several algorithms are available to assist with the analysis of developmental suitability, and most have been extended to take advantage of the map-data processing capabilities of a geographic information system. Six common methods have been reviewed extensively in the environmental planning literature: (1) the Gestalt method, (2) the method or ordinal combination, (3) the linear combination method, (4) the nonlinear combina-

Table 5.7 General methods for developmental suitability analysis.

Method	Description
Gestalt method	Determination of suitability classes through field observation, air photo interpretation, or topographic maps without consideration of individual environmental factors
Ordinal combination	Mapping the distribution of land types, subjectively rating their suitability, then physically overlaying each map to describe a composite suitability profile
Linear combination	Rating and weighting environmental factors, then applying an algorithm to produce a mathematical expression of "fitness"
Nonlinear combination	Use of a nonlinear function to combine ratings into a suitability score
Factor combination	Modification of the Gestalt method to accommodate interdependence among factors
Clustering	Application of statistical clustering algorithms to find natural grouping among environmental variables
Logical combination	The use of rule and heuristics to assign suitability scores to environmental factors

tion method, (5) the Rules of Combination method, and (6) statistical grouping techniques (Hopkins, 1977; Anderson, 1980; Lein, 1990). While the intent of each technique is to provide an objective measure of suitability, each of the procedures introduced tends to be based on a common set of assumptions. First is the assumption that landscape factors are independent. This implies that environmental parameters do not share functional relationships. We know, however, that landscape variables are greatly interdependent. Next is the assumption that mathematical properties hold when assigned values are compared across factor levels. Here, the implication is that adding or multiplying variables creates mathematically valid results. We understand, however, that three times soil does not equal slope. Another important assumption is that rating or scoring schemes can be applied uniformly across factor levels. Finally, it is generally assumed that subjectively derived weights are flexible given changing environmental conditions.

These assumptions, however, lack absolute validity. This fact tends to frustrate the "scientific" aspects of suitability assessment and implies that the majority of combinatorial techniques are invalid when viewed critically or insufficient if used as the sole means for expressing developmental constraint. Of course, if these or similar techniques are invalid or flawed, then why are they used in practice? There are several responses to this question. First, it is important to recognize that environmental planning is as much an art as it is a science. Therefore, while we may strive to develop analytic tools that embrace the rigor and objectivity of pure science, our reality is a blend of imprecise concepts coupled with environmental relations that defy exact quantification (Lein, 1993b). Consequently, techniques that perform composite landscape assessments should not be looked upon as a calculus of certainty, but rather as simplifying tools that reduce the complexity of the planning problem to a more manageable set of conditions. By keeping this point in mind, the planner may apply these techniques and use technical judgment effectively within its proper context. Suitability analysis, though conducted using addition, multiplication, or some other logical operation on a set of environmental conditions, is not an exercise in mathematics. Instead it characterizes an algebra involving symbol manipulation whose logic must be clear and explicit to ensure that its results are interpreted correctly. For this reason, developmental suitability analysis represents an expert judgment method of evaluation (McAllister, 1986). As an exercise in the guided use of expert judgment, the general methods used to express suitability and form a geographic pattern of landscape opportunity and constraint can be examined.

Gestalt methods

The Gestalt concept when applied to development suitability analysis requires consideration of the landscape as a whole rather than as an assemblage of elements. Homogeneous regions are determined through observation using data sources that characterize the landscape in its entirety, such as aerial photographs, topographic maps, or field observations. From these sources, an expression of suitability forms from a three-step process:

1 The study area is partitioned by implicit judgment into homogeneous land units based on the interpretation of landscape patterns.

2 A table is created that verbally explains the effects or constraints that will occur in each of the regions should the potential land use be located there. Each land unit is then evaluated and a relative score or grade is assigned.

3 A set of maps, one for each land use, is produced to show these regions in terms of their suitability. Value judgment is implicit and can be expressed graphically for each land unit delineated on the map.

The Gestalt technique yields valid or invalid analytic results depending on the skills of the analyst and the circumstances surrounding the project. Several problematic issues concerning this approach to suitability assessment have been noted. First is the realization that the process is based on implicit judgment rather than explicit rules. As a consequence, the results can be difficult to evaluate since the methodology is not well documented. Therefore, the procedure relies on one's mental ability to assimilate many interactive factors, which creates results that can be difficult to communicate to decision-makers.

Combinatorial methods

Generating an expression of suitability using combinatorial methods describes the general procedure of applying a mathematical operation directly or implicitly on a set of factor maps that depict important landscape qualities. Three principal methods dominate this approach: (1) ordinal combination, (2) linear combination, and (3) non-linear combination.

Ordinal combination – The ordination combination method explains a three-step procedure that creates a composite map detailing the developmental suitability of the planning region. The procedure begins by mapping a set of selected environmental variables according to specific categorical representations (i.e. soil types, slope classes, vegetation types). These mapped characteristics reflect distinct dimensions along which variations between land parcels can be described. Types mapped for each factor define nominal labels used to place variables along a measurement dimension. For example, land use might be typed according to well-recognized categories such as residential or industrial, while slope may be typed as low, moderate, high. The next step in the process requires creating a cross-tabulation table comparing factors to proposed or potential land uses. The elements of this simple two-dimensional matrix are filled by entries defining the relative suitability rating for each land use of each type across all the factors. The ratings explain an ordinal scaling of all the characteristics of the type. For instance soil type might include consideration of its permeability, depth to water table, organic content, and pH, together with the comparative "costs" of the land use it placed on the type. These ratings, expressed according to an ordinal measurement scale, can be derived from a number of sources such as maps, reports, or expert judgment. From here, a series of factor maps can be produced showing the suitability of the landscape for each land use under review. The final step in this process consists of physically overlaying the suitability maps of each individual factor for each land use to create a composite characterization of suitability over the region. If the number of factors is few, the visual interpretation of the composite map is relatively straightforward. However, as the number of factors increases, the associative logic needed to draw effective interpretations and recognize patterns in the data can confound meaningful interpretations. The greatest limitation of this approach, however, surrounds the implied addition of ordinal scale numbers and the assumed independence of factors. These flaws notwithstanding, the need to express suitability as a composite quality or score that will evidence spatial variation and can be displayed geographically over the planning region is clearly demonstrated by this approach.

Linear combination – The logical flaws inherent in the ordinal combination method are avoided when the types with each factor are rated on separate interval scales. The ratings for each type produced in this manner are typically multiplied by a weight to reflect the relative importance of that factor. The new "weighted" rating for a given area is then added to produce a suitability score. According to this approach, suitability becomes the product of the linear combination of factors. As demonstrated by Hopkins (1977), multiplication by weights can also be used to change the unit of measure of the ratings by the ratio of the multiplier, so that all ratings fall along the same interval scale.

Perhaps the most critical aspect of the linear combination method considers the procedures used to rate factors. Although there are no formal rules to guide the process, a logic is assumed that aims to produce a rating that can be interpreted meaningfully. Several schemes for deriving ratings are described by Largo (2001) and Hopkins (1977). A familiar rating scheme follows the analogy of assigning grades to a set of examinations. The first step in this process requires establishing a total possible suitability score. This value is divided among various factors and each assumes a given proportion of the total. Each type (class) of each factor is then rated as to its suitability in relation to the proportion of the total score assigned to the factor. Although the linear combination method corrects the measurement problems associated with ordinal approaches, the problem of factor independence remains. Linear combination methods are unable to address the situation where the relative suitability for a specific land use of a type on one factor depends on the type on any of the other factors. Despite this drawback, applying the linear combination method can be justified on the grounds that:

- Factors might be known *a priori* to be independent.
- The method is cost effective and easy to implement.
- Factors typically used in the model can be deductively determined to be independent.

Grouping techniques

An alternative strategy to suitability classification that avoids issues surrounding factor independence and mathematical invalidity involves the use of the methods that define homogeneous regions explicitly. These techniques apply multivariate statistical procedures to group or cluster land units into regions based on measures of similarity (Betters & Rubingh, 1978; Omi et al., 1979; Cifuentes et al., 1995). Three multivariate techniques have been shown to be particularly useful: (1) cluster analysis, (2) discriminant analysis, and (3) factor analysis/principal components analysis. When compared to judgment-based methods of suitability analysis, multivariate techniques provide a rational framework for analyzing the similarity of units that vary with respect to numerous environmental conditions. In addition, these techniques rely on a numerical treatment of data that implies direct measurement across several dimensions. This type of approach lends itself to more robust ordination and produces results that can be easily represented in map form.

As a procedure for determining developmental suitability, multivariate methods require certain information to guide the interpretation of results:

1. Criteria to assess suitability – given a set of possible uses, the criteria needed to adequately express suitability must be specified. These criteria may involve selected environmental characteristics as well as other factors considered relevant to the problem.

2. Criteria to determine importance – the criteria developed to assess suitability will vary in importance depending on the land use under consideration. These contrasts and variations in importance should be examined across all land uses.

3. Conditions that determine favorability for a given use – given different land uses, the conditions of favorability may vary for each criterion, therefore each criterion may have a somewhat different set of conditions considered advantageous or adverse to each use.

4 Suitability classification scheme – a classification system should be developed in accordance with the specified criteria. Above all, the method of classification should be objective and provide for a hierarchical characterization of suitability.

It is with this final point, classification, that multivariate techniques make their contribution.

Suitability by clustering Cluster analysis defines a set of techniques for accomplishing the task of partitioning a set of objects into relatively homogeneous subsets based on inter-object similarities. There are four strategies employed in a clustering procedure:

1 Partitioning methods
2 Arbitrary origin methods
3 Mutual similarity procedures
4 Hierarchical clustering methods.

Although a discussion of each is beyond the scope of this section, a detailed explanation of cluster analysis can be found in Davis (1986) and Kachigan (1986).

Typically, clustering begins by measuring each of a set of *n*-objects on a set of variables (*k*). Next, a measure of similarity, distance, or difference is calculated and used to compare objects in the set. Using these calculated similarity measures, an algorithm expressing a specific sequence of rules is employed to cluster the objects into groups based on the inter-object similarities. The logic underlying a given clustering algorithm is based on the assumption that objects that are similar belong to the same group. The ultimate goal of any clustering method is to arrive at a cluster pattern of objects that displays small within-cluster variation, but large between-cluster variation (Kachigan, 1986). The differences between the resultant clusters (groups) can be interpreted by comparing each to their mean values on the input variables. The two main considerations that influence the application of this technique and the selection of a specific clustering procedure relate to: (1) obtaining a measure of inter-object similarity, and (2) specifying a procedure for forming the clusters based on the similarity measures.

An expression of suitability develops out of the formation of the clusters as land units are grouped into similarity classes. This pattern is refined as the clusters are examined and the statistical groups are assigned to informational categories. The interpretation of the cluster pattern and the naming of the resulting groups is a critical phase in this application. In general, interpretation relies on the use of judgment supported by the careful examination of the means and variances on the environmental variables used as input. For example, at the conclusion of an analysis it might be observed that Cluster 1 displays a high mean on variables x_2, x_3, and x_7, while Cluster 2 shows high mean values on variables x_1, x_5, x_6. If variables x_2, x_3, and x_7 explain soil characteristics, then Cluster 1 may reflect soil qualities or conditions. Therefore, from these comparative profiles of the input variables the distinguishing qualities of the clusters can be identified.

Suitability by discriminant analysis Discriminant analysis is a procedure for identifying relationships between qualitative criterion variables and quantitative predictor variables. The technique is particularly useful for identifying the boundaries between groups of objects, where the boundary is defined in terms of the characteristics of the criterion variables that distinguish (discriminate) the object in the respective criterion groups.

Discriminant analysis is an adaptation of regression analysis and is designed for situations where the criterion variable is qualitative. In regression analysis an equation is solved that describes a weighted combination of values on various predictor variables. This equation enables prediction of an object's value on a quantitative criterion variable given its measure on each of the predictor variables. A similar concept is employed in discriminant analysis, only here the equation is called a discriminant function. The discriminant function uses a weighted combination of predictor variables to classify an object into one of the criterion variable groups. This function is therefore a derived variable defined as the weighted sum of values on the individual predictor measurement. This derived variable is termed a discriminant score.

The general form of the discriminant function can be expressed as:

$$\lambda = \beta_1 X_1 + \beta_2 X_2 + \ldots \beta_n X_n, \qquad (5.2)$$

where $X_1, X_2, \ldots, X_n$ represent values on various predictor variables, while $\beta_1, \beta_2, \ldots \beta_n$ explain weights associated with each variable. The value λ defines an object's discriminant score. Central to the application of this technique are the defining characteristics or parameters that guide its use. These include: (1) the weights associated with each predictor variable, and (2) the critical cut-off score for assigning objects into their criterion groups. Typically, these parameters are determined in such a way as to minimize the number of classification errors.

Applying discriminant analysis involves the following steps:

1 Selecting from a tentative list of variables those that contain the most useful classificatory information.
2 Constructing the relevant discriminant functions or scores based on the selected variables.
3 Interpreting the discriminant functions in order to identify factors that reflect major group differences.
4 Examining the discriminant functions in order to study the effects of several variables on an individual's group identity.
5 Making the final classification and validating the results.

Suitability by factor methods Factor analysis describes a family of procedures for removing the redundancy from a set of correlated variables and representing these variables as a smaller set of "derived" variables or factors. Perhaps one of the more applicable derivations of this approach is the technique referred to as "principal components analysis."

Although it is not strictly a method of factor analysis, principal component analysis (PCA) is a multivariate statistical procedure used to determine the underlying dimensionality of a data set. The main objectives of the PCA transform are to:

1 Reduce the dimensionality of a data set.
2 Determine a linear combination of variables.

3 Direct the choice of meaningful variables in an analysis.
4 Assist visualization of multidimensional data.
5 Identify underlying or latent structures in data.

In the majority of applications, PCA is employed to reduce the dimensionality of a data set and produce a set of components that are uncorrelated and ordered in relation to the variance explained by the original data. Following this procedure, environmental data describing important landscape qualities can be represented as a matrix of the form:

$$F_n = \begin{bmatrix} f_{1n} \cdots f_{1p} \\ \vdots \qquad \vdots \\ f_{mn} \cdots f_{mnp} \end{bmatrix}$$

Where f_{1n} represent observations $1 \ldots m$ with attributes $n \ldots p$.

The matrix F_n can be linearly transformed to contain a set of eigen vector components $V1$. The resulting new variables, termed principal components, can be expressed according to the relation:

$$V_{ij} = a_{ij} X_1 + a_{ij} X_2 + \ldots a_{ij} X_n. \qquad (5.3)$$

When the input data explain geographic qualities or quantities, the components express composite spatial patterns that are defined in terms of component loadings. These loadings quantify the relationship each component shares with the underlying patterns found within the original data. The principal component transform has several characteristics that are of special interest when applied to the problem of suitability analysis. First, the total variance is preserved in the transformation. Second, the transform minimizes the mean square approximation of error. Lastly, this is the only transform that generates uncorrelated coefficients.

An important issue that influences the usefulness of PCA relates to the problem of component structure and interpretation. After the reduction of observation space has been accomplished, which initial variables contribute the greatest share in the variance is often difficult to determine. Equally difficult to explain is the overall meaning of the component patterns. Thus, regardless of the application area, it is generally impossible to specify beforehand the number or significance of the components that will emerge from PCA. Consequently, since PCA may be regarded more as an exercise in mathematical manipulation than a pure statistical procedure, its value is best judged by performance rather than by theoretical considerations. Therefore, interpretation of the resulting components, their structure and meaning relies on judgment, the use of supporting information, and careful validation.

Extensions based on artificial intelligence

The suitability analysis problem has attracted interest from a variety of fields beyond environmental planning. Several of the more promising methodological approaches to assessment have emerged from the field of artificial intelligence (Lein, 1997). One AI technology in particular has been shown to be a viable alternative to traditional methods of suitability analysis (Lein, 1990). This alternative identifies a branch of AI research referred to as "expert systems." An expert system is a computer program designed to mimic the reasoning process of a human expert in a narrowly defined subject area. Using this expert knowledge, the system can approach problems using symbolic forms of data and information processing and reach conclusions similar to those a human expert would reach. In a paper by Lein (1990), a rule-based expert system was developed and applied to the suitability assessment problem. The system was designed to incorporate the subjective-technical judgment common to the procedures discussed previously, and treat them explicitly in the assessment process. The prototype system functions as a consultative expert system and uses a backward chaining inferencing strategy. The demonstration program consists of 110 rules, 223 qualifiers, and 45 choices that guide the planner through an assessment of suitability.

The primary intent of the prototype is to assist the planner in identifying relevant information and to direct the interpretation of effects. The user of this system interacts with a knowledge-base

Table 5.8 Sample rule in knowledge-based suitability system.

IF:	
	Proposed use is single family dwelling
	AND septic system is YES
	AND soil_permeability is 2.0 to 0.2 inches/hour
THEN:	
	Soil_limitations are moderate

Table 5.9 Sample conclusion window of suitability limitations.

Values based on 0.0 to 10.0 system	Value
1 flood limitations are. . . . **Severe**	10.0
2 groundwater limitations are. . . . **Severe**	10.0
3 bedrock limitations are. . . . **Moderate**	9.0
4 slope limitations are. . . . **Moderate**	9.0
5 drainage limitations are. . . . **Slight**	8.0

made up of rules and facts describing the suitability of selected environmental characteristics relative to a set of potential users (Table 5.8). By responding to a series of system queries, information is returned to the user listing the alternatives that best match the environmental conditions of the site (Table 5.9).

When compared to traditional methods of assessment and the limitations associated with each, the expert system becomes an attractive alternative that facilitates the synthesis of facts and qualitative experience into an automated support tool that can:

- Help clarify knowledge and effective problem-solving strategies.
- Preserve knowledge and encourage its sharing.
- Integrate knowledge and experience from several different fields.
- Apply heuristic knowledge.

A related AI technology that has been applied to the suitability analysis problems is the artificial neural network (Wang, 1994). Based on an abstraction of the human brain, an artificial neural network is a mathematical model comprised of a series of highly interconnected computational elements linked together to form a specific architecture. From this structure, information is processed according to a set of external inputs (stimuli). To demonstrate the feasibility of this approach, Wang (1994) developed a back-propagation neural network that takes as input a series of environmental variables then maps the input data to one of four suitability classes. Because neural network models are powerful pattern classifiers, the are extremely useful in situations that define high-dimensional problem spaces and complex interactions between variables. As illustrated by Wang (1994), a neural network can assess land suitability and provide classification accuracies of 80% or better. Therefore, when neural networks are compared to existing methods of analysis several advantages can be noted:

1 They can be trained to make decisions based on a more complex decision rule.
2 They are simple in structure and comparatively easy to construct.
3 They can be modified to fit other applications.

Carrying-capacity analysis

Carrying capacity has been a central concept in planning and environmental management for well over three decades (Mitchell, 1989). The concept emerged from the biological sciences, but when it is applied in the context of environmental planning, carrying capacity can be defined as the degree of human activity that a region can support at an acceptable quality of life without engendering significant environmental degradation (Bishop et al., 1974). Alternatively, Hayden (1975) and Cook (1972) have defined carrying capacity as the maximum ability of an environment to continuously provide resources at the level required by the population. Both definitions are valid and suggest that the interaction between population, development, and the local resource base is governed by thresholds or levels of intensity that will influence long-term sustainability.

Arriving at an expression of carrying capacity has traditionally relied on the subjective judgment of those familiar with the region in question, coupled with the measurement of surrogate estimators that could be related in a statistical algorithm

(Lein, 1993). Applying this expression, carrying capacity can be integrated with a range of social and economic indicators to define an "optimal" level of development. Based on this optimum, the environment may remain intact relative to its potential for sustained output. When one is considering the question of measurement, connecting the concept of carrying capacity back to its biological origins helps to frame an assessment methodology.

In the biological sciences, carrying capacity defines the relationship between the resource base, the assimilative and restorative capacity of the environment, and the biotic potential of a species. The biotic potential of a species, defining the maximum rate of population growth that could be achieved given the number of females that reach and survive through their reproductive years, is the controlling variable in this relationship. Biotic potential, given adequate food supplies, living area, and the absence of disease and predation, contributes to population increases that must be balanced by the environmental system. In the ecosystem, environmental resistance regulates biotic potential by imposing limits on food supply, space, and other inhibiting factors. Within this relationship, carrying capacity emerges as the limit or level a species population size attains given the environmental resistance indigenous to its location.

Whether it is explained within a biological context or from the perspective of a planner, carrying capacity is influenced by three critical assumptions:

1 There are limits to the amount of growth and development that the natural environment can absorb without threatening environmental stability through environmental degradation.
2 Critical population thresholds can be identified beyond which continued growth will trigger the deterioration of important natural resources.
3 The natural capacity of a resource to absorb growth is not fixed.

These assumptions have led to numerous attacks on the concept and the methodologies used to derive estimates of carrying capacity. Criticism tends to focus on questions concerning the definition of threshold levels, the representation of environmental equilibrium, and the manner by which carrying capacity estimates are interpreted (Lein, 1993a). These concerns have been summarized by Hassan (1982), and recognize the fact that (1) it is difficult to calculate environmental thresholds, (2) external influences can frustrate the "closed-system" perspective and introduce uncertainty, (3) strict concern for environmental potential limits the scope of analysis when applied to human systems, and (4) assessment relies heavily on subjective-professional judgment.

While the practical problems listed above conspire to frustrate the simple estimation of carrying capacity, the concept remains a useful applied methodology largely for its facility for:

- Raising questions about development strategies.
- Providing insight regarding the relationship between environmental degradation and human activities.
- Assisting in setting priorities in response to growth pressures.

Therefore, as Lein (1993a) maintains, the philosophical appeal and heuristic value of the carrying-capacity concept compensates for the pragmatic difficulties encountered with its application.

The majority of methods designed to assess carrying capacity (C) employ a deterministic solution to the general relation:

$$C = f(\text{potential resources, technology}).$$

Two central conditions are implied by this relation and influence the manner by which carrying capacity estimates are derived: (1) identification of a growth variable, and (2) identification of a limiting factor. A growth variable can represent either population or a measure of human activity such as the number of new dwelling units per year or the number of park visits per day (Ortolano, 1984). Limiting factors may include natural resources, physical infrastructure, or other finite elements that may restrain growth as a function of available technology. Limiting factors applied in carrying-capacity assessments help define three general expressions of the concept and direct its application

to problems where such measures can provide meaningful information. Three useful expressions of carrying capacity are:

1 **Environmental carrying capacity** – defined by biophysical characteristics (variables) including measures of air and water quality, ecosystem stability, and soil erosion. These variables define thresholds such as emission standards, BOD, and net primary productivity, that can be measured and linked by theory or empirical evidence to specific consequences. Employing these measures allows careful examination of the assimilative capacity of the environment and the ability of environmental "sinks" to accommodate change.

2 **Physical carrying capacity** – describing the capacity of infrastructure such as roads, highways, water supply systems, landfills, etc., to maintain an acceptable level of performance under population growth and development pressures. Because physical infrastructure is designed with specific capacity levels in mind, growth forces can exceed the predetermined optima and result in a degradation of performance and environmental quality.

3 **Psychological carrying capacity** – directs focus to the social environment and explains the role perception, attitude, behavior, and culture play in the way people react to their surroundings. Embedded in this expression are cultural and psychological factors that influence individual behavior and responses to the quality and condition of amenity resources, recreational area, institutional settings, and the aesthetic aspects of the environment.

Making carrying capacity work

Admittedly, carrying capacity, while it is intuitively appealing, is difficult to implement. However, it can be used to help formulate plausible alternative scenarios for how a region may develop or to refine a basic understanding of possible environmental constraints. To operationalize an exercise in carrying capacity analysis it is necessary to:

1 Identify the relevant growth variables and limiting factors.
2 Establish minimum and maximum values for each limiting factor.
3 Derive a quantitative link between limiting factors and growth variables based on either mathematical models, empirical relationships, or expert judgment.
4 Develop growth scenarios to explain (characterize) the dynamics of the processes involved.
5 Estimate the restrictions imposed by each limiting factor on the growth variable(s).
6 Review carrying-capacity estimates and refine methods as needed until results communicate constraint effectively.

Although the procedures for applying the concept to environmental planning are still evolving, Ortolano (1984), Mitchell (1989), and Westman (1985) have reviewed several useful examples.

Summary

Before decisions can be made regarding the future state of the planning area, information pertaining to the region must be collected, assembled, and analyzed. Because planning is an information-driven activity, methods to manage information and direct its use are critical to successful plans. In this chapter the collection and analysis of landscape information was examined. Drawing on the basic principles of regional landscape analysis and the formulation of a natural resource inventory, this chapter described how land evaluation methods can be applied by the environmental planner and how this information can be used to balance human development with the large goal of maintaining environmental functioning. The principal methods examined were land capability, developmental suitability, and carrying-capacity analysis; and the ability of each to provide an expression of landscape opportunity or constraint was evaluated.

Focusing questions

What does it mean to inventory a site and what does an inventory communicate?

Define the process of land evaluation and discuss the two basic types of land evaluation method.

What is the ultimate goal of developmental suitability assessment methods and what are the limitations associated with rating and weighting techniques?

How do the terms suitability and capability differ, and what does this difference mean?

CHAPTER 6

Natural Hazard Assessment

Critical to the responsibilities of the planner is the provision of a safe and healthful environment. This particular responsibility directs attention in part to efforts designed to minimize risk and reduce vulnerability to natural hazards. In Chapter 4 we briefly examined a set of natural processes that by virtue of their incidence describe hazards to human populations that share a common geographic location. These processes, while deemed hazards by people, have always existed, and simply define events that shape the earth's surface and identify earth-system responses to the dynamics of climate, geology, and hydrology that explain the realities of a living planet. Therefore, the environment is not hazardous as much as it is a challenge to human populations that seek to use its resources where the presence of certain natural events introduces an element of risk to life and property. From this perspective we can examine the nature and distribution of natural hazards, recognize the risks associated with each, and explore strategies to avoid risk and minimize the threat to human life and property.

The concept of planning with hazard is a somewhat novel approach when compared to the typical mitigation and control strategies that have been employed in the past. Yet, it is an important departure that recognizes the simple fact that rivers flood, hurricanes form in the mid-Atlantic and track toward coastal areas, and some forests do need to burn. This observation also follows the realization that many natural events when subject to human control strategies are made many magnitudes more severe as a consequence of our management efforts. The alternative paradigm of planning with nature rather than attempting to tame it encourages sustainable solutions to hazard management and more realistic programs to reduce vulnerability (Burby, 1998). In this chapter we will examine the nature of hazard and explain the relationship between natural hazards and environmental planning.

Defining hazards and risk

The terms hazard and risk are applied in close association but are often used imprecisely with different implicit meanings. A hazard can be defined as the potential to cause an adverse effect that can lead to harm. Extending this basic definition to include natural events, a natural hazard represents the potential interaction between humans and those natural events that are extreme in nature (Tobin & Montz, 1997). The hazard represents the potential, not the event itself, and this distinction is important because it recognizes the probabilistic nature of hazards and the uncertainty that underlies both the concept and the generating events that introduce it to a population. Therefore, by definition, natural hazards constitute an ever-present threat to society and represent an intrinsic force with which all societies must cope to some degree (Tobin & Montz, 1997).

The hazard exists largely because humans or their activities are constantly exposed to these natural forces, whether on the coastline of North Carolina or along the Hayward Fault on the eastern shore of the San Francisco Bay. Once an event occurs it is referred to as a natural disaster, where the term disaster is defined in relation to the significance of its impact on society. The difference between a hazard and a disaster is therefore made on the basis of causality. We conceive disaster owing to its physical form and geographic consequence, but the potential is ever-present.

Risk may be defined as either the occurrence of an event that results in harm, or as the probability of that occurrence and its consequence. Risk is therefore a presence that can be characterized quantitatively or qualitatively in terms of its likelihood, and can be conceptualized in two fundamental ways (Whyte & Burton, 1980):

1 **Risk as hazard** – a perspective that considers risk as synonymous with hazard: an event or act that holds adverse consequence where the degree of risk is related both to its probability and to the magnitude of its effects.
2 **Risk as probability** – a perspective that explains risk according to probabilistic statements or models: risk defines the probability value of an undesirable event.

Accordingly, reasoning about risk has developed following two contrasting approaches: the normative model of scientific reasoning based on probability theory, and a human-based approach drawing from symbolic logic and symbol manipulation (Lein, 1992). Based on the scientific-reasoning model, risk develops from the probability of a consequence (A) given that event (B) has occurred. This relationship may be expressed as the conditional probability (p) where:

$$p(A/B) = \frac{p(A) \cdot p(B/A)}{p(B)}, \qquad (6.1)$$

or in the case of logic stress analysis, a Bayesian probability drawn from a sequence of events of the form:

$$p(A/B) = \frac{p(A) \times p(B/A)}{\Sigma p(B_n) \times p(A/B_n)}. \qquad (6.2)$$

In other instances, risk may be expressed using mathematical expectation to predict the likelihood of an event (E) from a set of variables (X), such that

$$E = a_i X_i + a_j X_j + \ldots a_n X_n. \qquad (6.3)$$

Human reasoning, by contrast, tends to follow a less rigorous model that is based largely on the use of vague or imprecise concepts that take meaning only within the context of language (Lein, 1992). Risk reasoning in this example becomes a function of individual knowledge, behavior, and prior experience. As demonstrated by Klein and Methlie (1990), the theory of human reasoning suggests that people apply very few formal principles when considering risk or any other type of problem situation. Instead, people rely on heuristics (rules of thumb) tailored to the semantic context of the problem. When one is considering risk, problems often develop when normative models of probability cannot be used to drive meaningful estimates or when such estimates must be interpreted by decision-makers who do not reason in terms of mathematical probabilities. As a consequence, it often becomes necessary to distinguish between empirically based findings of risk and those derived from judgment. This need is becoming increasingly apparent, particularly in situations where risk characterization relies on expert judgment; a situation amply demonstrated by Flemming (1991) and Bonano et al. (1989).

The relationship between hazard and risk is further complicated by the difficulty in separating effects from their actual causes. Hazards have three important components that function in concert to punctuate their significance. These hazard components include a physical perspective, a human dimension, and a spatiotemporal disposition (Tobin & Montz, 1997).

The physical dimension directs attention to the geophysical world and the processes that conspire to define hazards. From this perspective, the physical world is seen as an external force separate from the human world. Although the traditional idea that natural hazards are the exclusive result of geophysical processes has been replaced by a more considered view, knowledge of physical process remains a critical aspect of hazard identification. A moderating influence, that has tempered the conceptualization of hazard and defines process as those external forces that act on human populations, has come from attempts to produce human explanations of natural hazards. This area defines the human dimension component of the hazard and directs focus toward the interaction between physical process and human forces that in combination determine the significance of disasters and produce a more realistic description of risk. For example, Smith (1992) suggests that natural hazards result from a conflict between geophysical processes and people. Hazards, according to this view, develop at the interface between the natural event and the human-use system. In some circumstances the focus of concern eliminates physical processes entirely and concentrates attention on the disruption to society following hazard events. The human dimension considers how events interfere with everyday life, disrupt communities, strain local services and resources, and leave an aftermath that can persist long beyond the initial event. Consideration of a hazard's human dimension must therefore evaluate not just the primary consequence of an event, but the secondary and tertiary effects, and beyond, that continue to shape the human-use system. Recognized in this description are those events that can be triggered by human activities and consequences that can include economic, social, psychological, and technological factors.

While it is accepted that when one is characterizing natural hazards, both physical processes and the human mosaic are important elements of the risk equation, an equally critical factor in defining hazard and risk is the timing of the event and its geographic extent (expression). Time and space have a range of influences on the nature of hazard and the definition of risk. For example, when one is considering a natural hazard, time may be expressed in relation to an event's:

- Frequency
- Duration
- Seasonality
- Timing.

Each of these temporal characteristics must be defined in relative terms, with sensitivity to their conceptual overlap. When placed into the context of a specific process or event they help punctuate where and how in time events manifest. For example, declines in rainfall may have occurred well before drought is identified; similarly, a severe thunderstorm with damaging hail may pass through the planning area in a matter of minutes, only to be followed by another storm complex the next day. Contrasted to this pattern are processes such as earthquakes above a certain magnitude that have reoccurrence intervals of hundreds to thousands of years. Therefore, connected with each hazard is an implicit expression of time. These time traits complicate how society responds to hazard and, more importantly, how risks are perceived and understood.

Because natural hazards form out of natural processes they define a unique geographic expression that places their origin and impact into a definable spatial context. The geographic nature of hazard assumes two critical explanations that must be understood by the planner: (1) location and (2) scale. Some natural processes are spatially ubiquitous in that they operate nearly everywhere on the planet to some degree. Extremes of heat, dryness, and wind velocity may be examples of geophysical processes that are not specific to a fixed geographic location. Conversely, certain natural processes are confined to specific geographic areas or zones, and exhibit a higher frequency of occurrence in particular locales. Thus fault movements are geographically restricted to regions were tectonic processes are active, tornadic activity is more frequent in climatic zones where conditions favor their development, landslides can be anticipated in areas where slope, soil, and geologic conditions are unstable and induce mass wasting. These illustrative examples suggest that hazards display a geographic patterning, and link to geographical processes that share a discernible regionality. Geography also conspires to influence the spatial scale or extent of the hazard. Scale, when considered in relative terms, defines the size of the geographic area where the event and its consequence will form. Some natural events may be highly localized and confined to a limited geographic area. Other events may occur on a scale capable of affecting a much broader geographic area. It is possible to even consider the global effects of some hazards as their influence reaches well beyond where they physically occur. A drought, for example, may influence a large area, perhaps at the continental scale, while the implications of that drought on agriculture may have a global impact. Treating scale as an elastic concept, it can be seen that areas from several square miles to portions of entire continents can be subject to extreme processes that can introduce hazard and place large populations at risk. It is critical for effective planning to understand the meaning of hazard and to place risk into its proper context. A complete understanding, however, is not possible without consideration of those physical processes that induce hazardous events.

A typology of hazard

Natural processes are neutral in their disposition toward human populations; however, when these processes contribute to injury, death, or damage to property they are defined as hazards. But we also recognize that human activities may alter the frequency with which these events occur, increase or decrease their severity, alter the size of the area impacted, influence the rate of exposure of people or property, and influence the vulnerability of hazard-exposed populations (Petak & Atkisson, 1982). For the purposes of rapid identification, natural hazards can be defined in relation to the underlying physical processes responsible for their genesis. These driving natural processes can be placed into three categories: (1) meteorological, (2) hydrological, and (3) geological. Within these general categories hazards can be discussed according to their

- Physical mechanism
- Temporal distribution
- Spatial distribution
- Onset pattern.

Meteorological processes

Extreme meteorological events develop when the factors that define day-to-day weather exceed critical thresholds. The typical elements of weather, such as temperature, wind speed, humidity, radiation, and precipitation, become important when the patterns they form characterize potentially problematic situations. Several of the more common events of weather that achieve hazard status include:

- Tornadoes
- Tropical cyclones
- Severe storms
- Extremes of cold and heat.

Tornadoes A tornado can be described as a violently rotating column of air, a vortex spawned by a thunderstorm, in contact with both the thunder cloud and the ground, often accompanied by a funnel-shaped cloud, progressing over the land in a narrow path (Grazulis, 1993). Recognized as among the most powerful of weather phenomena, a tornado can produce rotating velocities approaching 500 mph and affect a ground area ranging from $1/4$ to $3/4$ miles wide. The path a tornado follows can be as short as a few tens of yards to over 15 miles long. Extreme events have been known to travel over areas measuring up to one mile wide and 300 miles long (Petak & Atkisson, 1982). Although a tornado can occur anywhere severe thunderstorms develop, certain geographic areas evince climatic regimes that foster more frequent tornado development. In the United States the areas comprising the mid-western and southeastern regions display a higher frequency of tornadoes and tend to be areas of heightened vulnerability. Although tornadoes can occur at any time throughout the year, in North America weather conditions between April and August are the seasonal maximum for these events.

Generally, tornadoes form in association with squall lines with both isolated thunderstorms and thunderstorms accompanying frontal passages. As the storm system develops, one or more tornadoes may form at intervals along the storm track, travel a few miles, lift, and then reform further downfield. With an incidence time lasting minutes to hours, tornadoes reach much higher speeds than a hurricane, but affect a much smaller geographic area. To communicate the power and intensity of these systems several scales of measurement and classification have been devised.

Table 6.1 The Fujita–Pearson tornado scale.

Scale	Wind speed	Effects
F-0	40–72 mph	Chimney damage, tree branches broken
F-1	73–112 mph	Mobile homes pushed off foundation or overturned
F-2	113–157 mph	Considerable damage, trees uprooted
F-3	158–205 mph	Roofs and walls torn down, cars thrown
F-4	207–260 mph	Well-constructed walls leveled
F-5	261–318 mph	Homes lifted off foundation and carried considerable distances, autos thrown near 100 meters

One of the more widely adopted scaling methods is the Fujita classification system (Table 6.1). The Fujita classification system assigns a numeric value to a tornado based on its wind speed in $1/4$ mph increments. The Pearson scale is another complementary classification system for categorizing tornado events. This scale applies a logic similar to that used by the Fujita method, but the Pearson scale classifies tornadoes based on path width and length. Using these scales, the magnitude of hazard can be expressed and that information can be used to devise adjustment strategies for tornado events. For example, for a given structure, the expected damage from tornado exposure increases as a function of the Fujita rating of occurrence. With an event given a Fujita rating of 5 (F5), we could anticipate that 65% of all exposed woodframe structures would collapse. This would compare with an F1 event where only 1% of structures would collapse. Perhaps the most vexing problems associated with tornadoes from a planner's perspective are that:

1 While conditions under which they form are understood, the mechanisms that cause them to form remain unclear.
2 They are extremely difficult to measure directly.
3 Mitigating the effects of tornadoes is frustrated by the fact that these events, while localized, are highly random.

Tropical cyclones　More commonly known as hurricanes, tropical cyclones involve a mix of devastating winds, flood producing rains, and potentially lethal storm surges. A hurricane is an intense storm of tropical origin with sustained winds exceeding 74 mph. They form over tropical waters where the winds are light, the humidity is high, and the surface water temperature over a wide surface area is warm (79°F/26°C) (Ahrens, 1991). Over the tropical and subtropical north Atlantic and north Pacific oceans these conditions prevail in summer and early fall. This corresponds with the seasonal peak of hurricane frequency during the months between June and November.

In general, several necessary conditions must be present for hurricanes to develop (Pielke, 1990):

1 An ocean area with surface water temperatures above 26°C (79°F).
2 Small wind speed and direction changes between the lower and upper troposphere of less than 15 km/h.
3 The presence of a pre-existing region of lower tropospheric horizontal wind convergence (a tropical wave).
4 A distribution of temperature with height which will overturn when saturated, resulting in cumulonimbus clouds.
5 A location at least 4 to 5 degrees away from the equator.

Hurricanes transition through a series of stages marking their birth to death. Initially a mass of thunderstorms with only a slight wind circulation develops into a tropical disturbance. When sustained winds increase to between 20 and 34 knots and a centralized trough of low pressure appears, the disturbance becomes a tropical depression. As the pressure gradient intensifies and wind speeds increase to between 35 and 64 knots, the tropical depression becomes a tropical storm. Once wind speeds exceed 64 knots (120 km/h) the tropical storm is classified as a hurricane.

Hurricanes that develop over the north Pacific and north Atlantic are directed by an easterly air-

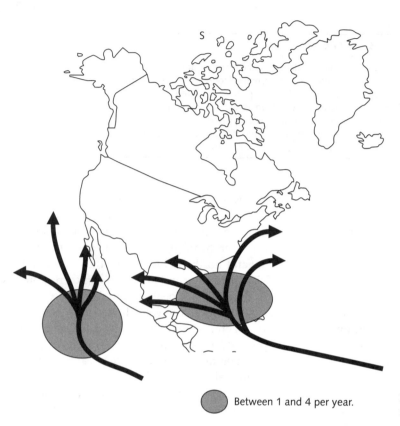

Between 1 and 4 per year.

Fig. 6.1 Typical hurricane pathways for North America.

flow. Gradually, these systems swing poleward around the subtropical high, and if they move far enough north, they are captured by the westerly circulation and are steered in a northerly direction. However, the path of a given hurricane can vary considerably. The typical pathways hurricanes follow are illustrated in Fig. 6.1.

While the high winds associated with a hurricane can inflict considerable damage, most of the destruction related to these events is caused by high seas, huge waves, and flooding. Flood risk is due partly to the force of wind pushing water onto the shore and to intense rainfall that can exceed 25 in (63 cm). The combined effect of high water and high winds produces a storm surge that inundates low-lying areas and easily overruns beach-front properties. The Saffir–Simpson scale was developed in an effort to estimate the possible damage a hurricane's sustained winds and storm surge might inflict on a coastal area (Ahrens, 1991). The damage assessment expressed by this scale is based on actual conditions observed during the life-cycle of the storm.

Severe storms Extratropical cyclones can produce a variety of hazards including hailstorms, severe winds, severe snows, and ice storms (Smith, 1992). Some mid-latitude cyclones create special hazards because they develop very quickly. These rapidly deepening depressions are often difficult to forecast since their rate of deepening is often underpredicted. In North America, particularly in the central and Great Lakes regions of the United States, ice and glaze storms are a significant winter hazard. In these geographic areas, the hazard arises when thick accretions of clear ice form on exposed surfaces. Ice accretes on any structure whenever there is liquid precipitation or cloud droplets, and both the air and the object's temperature are below freezing. When these events occur, electric power transmission lines, landscaping trees, and forests are at the greatest risk for damage. Here, the added weight of ice may be sufficient to bend or in some instances bring these objects down.

Temperature extremes   Periods of unusually cold or hot weather have been shown to cause a direct threat to human life (Smith, 1992). As suggested in the literature, the average human body is most efficient at a core temperature of 37°C. Given natural variations in temperature, physiological comfort and safety can be maintained within only a narrow thermal range. When the heat balance of the body deviates beyond this range, physiological stress results. Extremes of cold also create hazards in the form of ice and frost, while extremes of heat can be a contributing factor to elevated wildfire risks.

In a review of atmospheric hazards, Smith (1992) identifies several important considerations related to the effects of thermal extremes. First, with reference to physiological hazards, stress to the human body can result from a combination of low temperature and high wind speeds. This wind-chill hazard is common to high latitude and high altitude environments. However, the greatest threat is associated with unexpected outbursts of very cold air into the mid-latitudes in winter. Such a pattern occurs in North America with the development of a high-pressure ridge that forms over the northwestern margins of the continent. The circulation around this ridge allows arctic air to penetrate well into the mid-west and brings cold and frost conditions well into Florida.

Extremes of high temperature can also produce life-threatening situations by imposing severe heat stress on the human body. The problem can be particularly acute when high temperatures are combined with high relative humidity. When these situations arise, heat stress can be common, as the body's ability to cool through evaporative processes becomes less effective and efficient. In prolonged events the physical discomfort can quickly escalate into increased heat-related mortalities (Quayle & Doehring, 1981).

Atmospheric processes can also be modified by the changes in surface composition introduced by urbanization. These modifications have become recognized as the "urban climate," and while they may not represent hazards, they do explain changing conditions that may accentuate hazardous conditions. For example, the altered heat balance of the urban environment contributes to the formation of "hot" sectors in the urban landscape. In these sectors airflow is reduced, emissions concentrate near the ground, and evaporative cooling is substantially reduced by the absence of vegetation. Where building heights increase, more turbulent airflow is common, as is the reduction in direct solar radiation reaching the surface. In these urban canyons, cooler and windier conditions prevail. The general pattern of local climate influenced by urbanization is shown in Fig. 6.2.

Hydrological processes

Hydrologic processes are characterized by the flux parameters that define the hydrologic cycle. These transfer mechanisms influence the distribution of water at the surface and describe the general availability of water to the environmental system. When events transpire that move these parameters above or below their expected condition, a stress relationship can develop that can influence ecosystem functioning and the water resource systems on which human settlement depends. Perhaps two of the more common hydrologic events that evidence deviations sufficient to assume hazard status are floods and drought.

Mechanics of flooding   Physically a flood is a high flow of water which overtops either the natural or the artificial banks of a river channel (Smith, 1992). Under natural conditions, the hydrologic processes of the fluvial system have created and maintained stream channels with channel maintenance maximized during bank full discharges. Whenever flow exceeds bank full capacity, and the river spills over onto the adjacent floodplain, the river is at flood stage. All rivers produce this condition at some time, suggesting that flooding is a natural feature of river systems. Consequently, such events cannot be considered a hazard unless the nature of the flood threatens human life and property.

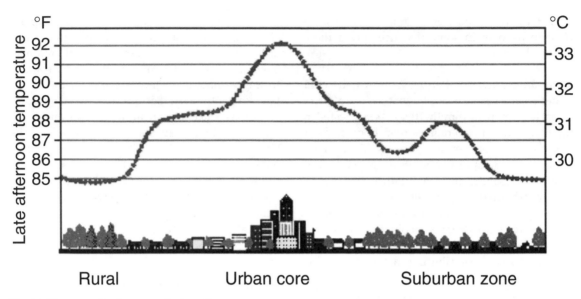

Fig. 6.2 The generalized pattern of urban climate.

When one is defining the concept of a flood, its physical cause can be expressed in two ways. One explanation stresses the hydrologic definition and defines flood magnitude in terms of peak river flow. This definition contrasts with a more hazard-oriented explanation. This alternate expression relates the flood event to the maximum height or stage that water reaches. These two definitions require a careful distinction between the primary causes of floods that result from the driving forces of climate, and those secondary flood-intensifying conditions that occur as a consequence of drainage basin morphology.

Perhaps the most important cause of floods is excessive rainfall. Rainfall events capable of producing floods can vary from seasonal storms that provide precipitation to a large geographic region to nearly random convectional storms that can generate flash-flood conditions over a comparatively localized area. In addition, some flood events can be causally linked to larger atmospheric processes such as the ENSO (el Niño) or its opposing process, la Niña. Within the mid-latitudes prolonged rainfall is frequently related to tropical cyclones or intense atmospheric depressions. Localized convectional systems are also responsible for producing high-intensity rainfall particularly during the summer season. These rainfall events, because they are temporally and geographically concentrated, can generate large volumes of water with a high damage potential. Although rainfall may be the driving force behind flood events, processes such as the seasonal melting of snow and ice can be contributing factors in some areas. The seasonal melt can create widespread flooding, particularly during years of high snow accumulation. Melting during these periods contributes to spring flood events that can be compounded by ice jams that may temporarily dam river channels.

When a flood occurs, the area of the stream's floodplain that will be inundated by water resulting from a stream flow level of a specified flood-frequency is referred to by the name of that frequency interval. This flood frequency interval and the geographic area it will encompass have tremendous importance to hazard assessment and planning. To illustrate this point, consider a flood characterized by a two-year return interval. Such an event will inundate all of the area delineated as the two-year zone of the floodplain. Similarly, a flood stage with a 100-year frequency will inundate the entire 100-year zone of the floodplain. The geographic expression and extent of these zones is depicted in Fig. 6.3.

N

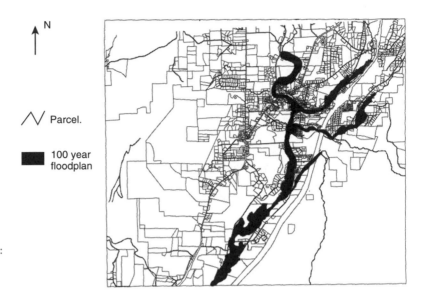

/\/ Parcel.

■ 100 year floodplan

Fig. 6.3 Delineation of flood zones: town of Manchester one hundred year flood hazard zone.

The flood frequency–floodplain relationship illustrated in Fig. 6.3 suggests spatial patterns that can greatly influence the present and future use of flood prone areas. For this reason, floodplains are typically divided into six flood classes or six hazard zones (Petak & Atkisson, 1982). The classification system categorizes the floodplain into zones A through F, where zone A defines the most hazardous area subject to the most frequent flooding, while zone F is the least hazardous. According to the logic of this classification system, the six flood/hazard zones delineate areas that would be inundated by the 2–5, 5–10, 10–25, 25–50, 50–100, and greater that 100-year flood event. Within the drainage basin, however, the geographic shape and extent of these zones will be influenced by local terrain characteristics and river morphology. Thus, the dimensions of a flood zone will differ, and the depth floodwater can reach given a specified magnitude will vary along the course of the river. Within these zones the damage sustained by structures is generally a function of:

1 The type, strength, and elevation of the structure.
2 The depth of the flood waters during the event.
3 The force exerted against the structure by moving floodwaters.

4 The impact of floating debris against the structure.
5 The adverse effects of water intrusion on the structure and its contents.

Frequently overlooked aspects of the flood hazard problem are those factors that contribute to flood-intensifying conditions, particularly those associated with human activities. Floods can be made more intense for a given precipitation level depending on topography, the hydraulic geometry of the drainage basin, soil characteristics, and the density of vegetation cover. In addition to these natural parameters, changes in land cover and land use are also contributing factors. In this regard, two of the more powerful forces that conspire to intensify flood conditions are urbanization and deforestation. Urbanization alters the magnitude and frequency of floods in several ways. Four well-documented impacts of urbanization on flooding are summarized in Table 6.2. Deforestation increases flood run-off with associated increases in surface erosion that decrease channel capacity through sediment deposition. The removal of vegetation causes peak flood flows to increase, and changes the hydrologic regime of the river basin, as evidenced by its unit hydrograph (Fig. 6.4).

Table 6.2 Impacts of urbanization on flooding.

Increased impervious cover
Increased run-off
Increased pollutant and sediment loads
Decreased vegetation cover

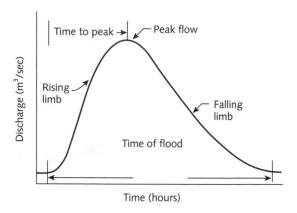

Fig. 6.4 Characteristics of a flood hydrograph.

Drought Drought is a complex hydroclimatic phenomenon and is perhaps the most persistent environmental hazard experienced in the contiguous United States (Soule, 1992). When compared to other environmental processes, drought is often referred to as a "creeping" hazard because it develops slowly over several months with prolonged effects often spanning a period of years (Smith, 1992). One of the more problematic issues related to drought is the question of definition. Generally drought can be explained in one of three ways (Drought Information Center, 1999), to which a fourth can be added.

1 Meteorological drought Meteorological drought is defined usually on the basis of the degree of dryness (in comparison to some "normal" or average amount) and the duration of the dry period. Definitions of meteorological drought must be considered as region specific since the atmospheric conditions that result in deficiencies of precipitation are highly variable from region to region. For example, some definitions of meteorological drought identify periods of drought on the basis of the number of days with precipitation less than some specified threshold. This measure is only appropriate for regions characterized by a year-round precipitation regime such as a tropical rainforest, humid subtropical climate, or humid mid-latitude climate. Locations such as Manaus, Brazil; New Orleans, USA; and London, England, are examples. Other climatic regimes are characterized by a seasonal rainfall pattern, such as the central US, northeast Brazil, west Africa, and northern Australia. Extended periods without rainfall are common in Omaha, Nebraska, Fortaleza, Brazil, and Darwin, Australia; a definition based on the number of days with precipitation less than some specified threshold is unrealistic in these cases. Other definitions may relate actual precipitation departures to average amounts on monthly, seasonal, or annual timescales.

2 Agricultural drought Agricultural drought links various characteristics of meteorological (or hydrological) drought to agricultural impacts, focusing on precipitation shortages, differences between actual and potential evapotranspiration, soil water deficits, reduced groundwater or reservoir levels, and so forth. Plant water demand depends on prevailing weather conditions, biological characteristics of the specific plant, its stage of growth, and the physical and biological properties of the soil. A good definition of agricultural drought should be able to account for the variable susceptibility of crops during different stages of crop development, from emergence to maturity. Deficient topsoil moisture at planting may hinder germination, leading to low plant populations per hectare and a reduction of final yield. However, if topsoil moisture is sufficient for early growth requirements, deficiencies in subsoil moisture at this early stage may not affect final yield if subsoil moisture is replenished as the growing season progresses or if rainfall meets plant water needs.

3 Hydrological drought Hydrological drought is associated with the effects of periods of precipitation (including snowfall) shortfall in surface or subsurface water supply (i.e., stream flow, reservoir and lake levels, groundwater). The frequency and severity of hydrological drought is often defined on a watershed or river-basin scale. Although all droughts originate with a deficiency of precipitation, hydrologists are more concerned with how this deficiency plays out through the hydrologic system. Hydrological droughts are usually out of phase with or lag the occurrence of meteorological and agricultural droughts. It takes longer for precipitation deficiencies to show up in components of the hydrological system such as soil moisture, stream flow, and groundwater and reservoir levels. As a result, impacts are out of phase with those in other economic sectors because different water use sectors depend on these sources for their water supply. For example, a precipitation deficiency may result in a rapid depletion of soil moisture that is almost immediately discernible to agriculturalists, but the impact of this deficiency on reservoir levels may not affect hydroelectric power production or recreational uses for many months. Also, water in hydrologic storage systems (e.g., reservoirs, rivers) is often used for multiple and competing purposes (e.g., flood control, irrigation, recreation, navigation, hydropower, wildlife habitat), further complicating the sequence and quantification of impacts. Competition for water in these storage systems escalates during drought and conflicts between water users increase significantly.

4 *Socioeconomic drought* Socioeconomic definitions of drought associate the supply and demand of some economic good with elements of meteorological, hydrological, and agricultural drought. It differs from the aforementioned types of drought because its occurrence depends on the time and space processes of supply and demand to identify or classify droughts. The supply of many economic goods, such as water, forage, food grains, fish, and hydroelectric power, depends on weather. Because of the natural variability of climate, water supply is ample in some years but unable to meet human and environmental needs in other years. Socioeconomic drought occurs when the demand for an economic good exceeds supply as a result of a weather-related shortfall in water supply. For example, in Uruguay in 1988–9,

drought resulted in significantly reduced hydroelectric power production because powerplants were dependent on stream flow rather than storage for power generation. Reducing hydroelectric power production required the government to convert to more expensive (imported) petroleum and stringent energy conservation measures to meet the nation's power needs.

In most instances, the demand for economic goods is increasing as a result of increasing population and per capita consumption. Supply may also increase because of improved production efficiency, technology, or the construction of reservoirs that increase surface-water storage capacity. If both supply and demand are increasing, the critical factor is the relative rate of change. Is demand increasing more rapidly than supply? If so, vulnerability and the incidence of drought may increase in the future as supply and demand trends converge.

For the purposes of environmental planning, the simplest conceptualization of drought defines it as an unusually dry period that results in a shortage of water (Smith, 1992). From this perspective deficiencies in precipitation may serve as the triggering effect; however, a careful distinction must be made between the shortage of precipitation and the shortage of useful water. Therefore, the concept of drought cannot be separated from the water resource and water allocation issues that surround water at its utilization. A precipitation deficiency associated with drought poses the potential to produce water supply problems, particularly in regions that rely on surface sources as the main origin of supply. Consequently the significance of drought and its relevance to the planning problem relates primarily to three controlling factors:

1 The purposes for which water is required.
2 The ways in which the local hydrologic cycles react to precipitation deficits.
3 The degree of buffering that is available to offset precipitation shortages.

When one is monitoring the duration and intensity of drought, it is convenient to apply an index that encapsulates the spatial pattern of the incidence of drought in terms of a single numerical rating (Katz & Glantz, 1986). A variety of indices

Table 6.3 Procedures followed for calculating PDSI.

Step 1 Estimate potential evapotranspiration, potential soil recharge, potential run-off, and potential loss for an average period of a month for each climatic subdivision.

Step 2 Use the month-by-month water balance accounting to obtain coefficient of evaporation, recharge, run-off, and loss.

Step 3 Compute the amount of precipitation that should have occurred during a given month to sustain potential evapotranspiration, run-off, and moisture storage that would be considered normal and climatically appropriate for existing conditions.

Step 4 Subtract the normal and climatically appropriate value from areally averaged precipitation to obtain a precipitation excess or deficit.

Step 5 Calculate a moisture anomaly index (Z) and determine the final drought index term from the general relation $Z/3.0$.

have been proposed and several of these have been compared in the drought literature (Oladipo, 1985). Two of the more widely applied drought indices in the United States are the Palmer Drought Severity Index (PDSI) and the Crop Moisture Index (CMI). Although both indices are accepted as representative measures of drought conditions, their focus is primarily directed toward meteorological definitions of drought.

The Palmer index (PDSI) provides an objective method for developing a meteorological characterization of drought. The index expresses drought as a function of precipitation, potential evapotranspiration, antecedent soil moisture, and run-off. Allen (1984) has reviewed the computational procedures followed to derive the index, and full details are given by Palmer (1965). The general procedure is summarized in Table 6.3. The crop moisture index (CMI) was developed primarily to assess and evaluate agricultural drought. The CMI calculates crop moisture conditions by considering the interrelationships between the deviations of precipitation levels from normal, soil-moisture supplies, and evapotranspiration demand.

Placing the derived drought index values into a classification scheme that groups drought strick-en areas according to severity or significance facilitates assessment of a drought's geographic pattern. Diaz (1983) has proposed a useful drought classification method. According to this system, a sequence of three or more months with PDSI values less than or equal to −2.0 is considered to represent a drought event, a period of six or more months a major drought event, while a mild drought is defined by PDSI values as less than or equal to −1.0.

Geologic processes

Geological processes relevant to a discussion of natural hazards typically describe events related to tectonic activity or the forces of erosion and mass wasting. When compared to other environmental hazards, the time-scales within which geologic processes behave frequently make them difficult to conceptualize in accurate terms. As a result, the energy released when these events occur is either underestimated or subject to alarmist reactions. Despite these contrasting modes of response, geologic processes introduce significant influences on the utilization of land resources and the density, design, and location of specific land uses. While the list of potentially significant geologic processes to consider is exhaustive, those germane to environmental planning include landslides and mass movements, subsidence and collapse, earthquakes and faulting, coastal processes, and volcanic activity.

Landslides and mass movements Mass downslope movements may occur in a wide variety of geologic materials. Soil, rock, or the combination of the two may fail or migrate downslope under a range of environmental conditions. Some slides may involve a comparatively small amount of material, while others may result from deep failures of large masses of solid rock. Movement can occur on very gradual slopes, on steep terrain, and under diverse climatic conditions. The slides and movements characterized above are part of a more general erosional process referred to as "mass wasting." This fundamental surficial process describes the downslope movement of earth surface

Table 6.4 Principal forms of downslope movement.

Soil or bedrock creep	The slow downslope movement or the gradual plastic deformation of the soil mantle at nearly imperceptible rates.
Rockfalls	The abrupt free-fall or downslope movement of rocks and loosened blocks or boulders of solid rock.
True landscapes	The failure of material at depth and the movement of that material along a rupture or slip surface.
Earth flows	The downslope movement of soil or overburden that become saturated by heavy rains.
Debris or mud flows	The rapid but viscous flow of mud or other surface material.
Snow avalanches	The rapid downslope movement of snow, ice, and associated debris such as rock and vegetation.

materials under the force of gravity. Mass movements can range from soil and rock creep, where rates of movement are measured in centimeters per year, to debris avalanches where velocities may reach 400 km/h. Landslides fall within the middle range of these extremes and can be identified by the presence of a surface rupture.

Various forms of downslope movement carry important hazard potential. Depending on the type of movement and the kind of material involved, mass movements can vary widely with respect to their shape, rate, spatial extent, and impact on the surrounding environment. Consequently, they can pose variable levels of hazard that planners must understand. Brief descriptions of those critical to the planning problem are given in Table 6.4.

To understand the landslide hazard it is important to recognize the factors that make an area susceptible to failure. Once the triggering mechanisms are identified it may be possible to initiate appropriate planning solutions to minimize the hazard and reduce risk. In general, mass downslope movements occur when the component of weight along a surface exceeds the frictional resistance or cohesion of the material (Griggs &

Gilchrist, 1977). Therefore, when the strength of the material that comprises the slope is overcome by a downslope stress, the slope fails. Two factors are central to this process: (1) shear strength which defines a slope's maximum resistance to failure, and (2) shear stress which is defined as the component of gravity that lies parallel to a potential or actual surface of slippage. When the equilibrium condition of a slope is disturbed, the shear stress on the material increases. The stress acting on a slope may be affected by several external and internal factors. Any of these may be sufficient to contribute to slope failure. External factors affect the stress acting on a slope and include:

- Changes in slope gradient
- Excess loading
- Changes in vegetative cover
- Shock and vibrations.

Internal factors alter the strength of the material comprising the slope and involve consideration of:

- Changes in water content
- Groundwater flow
- Weathering effects.

Subsidence and collapse Land subsidence is not typically considered a major geologic hazard, yet where it occurs it can pose serious problems to land use and infrastructure. While the process of land subsidence is gradual in most geographic settings, collapse can be sudden with catastrophic consequences. Seven major causes of land-surface subsidence and collapse have been identified. These include:

1 The withdrawal of large volumes of water, petroleum, or natural gas from weakly consolidated sediments.
2 The application of water to moisture-deficient deposits resting above the water table.
3 Tectonic activity.
4 The solution or leaching of soluble subsurface material by groundwater.
5 The removal of subsurface deposits of coal or other mineral resources with inadequate surface support.
6 The melting or disturbance of permafrost.
7 The differential settlement of artificial fill.

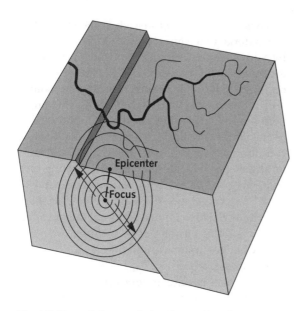

Fig. 6.5 General characteristics of an earthquake.

Earthquake and faulting Earthquakes may be characterized as natural events that involve the moving or shaking of the earth's crust. While the specific triggering mechanisms remain uncertain, it is believed that earthquakes are produced by the release of stresses accumulated as a result of rock rupture along opposing fault planes in the Earth's outer crust (Fig. 6.5). One common source of rupture results from the continuous collisions between plates that comprise the Earth's 10- to 50-mile-thick outer and floating crust. Three different mechanisms are associated with the constantly migrating patterns of these plates:

1 Divergence – describes the situation where tectonic plates spread, move apart, and increase in area.
2 Collision – results when plates converge and reduce in size, forming a subduction zone as one plate overrides another.
3 Transcursion – occurs when two plates slide laterally past each other, producing a series of tears or transcurrent faults.

As shown by Smith (1992), these tectonic processes give rise to a variety of primary and secondary seismic hazards (the secondary earthquake hazards include soil liquefaction, landslides and rockfalls, and tsunami and seiches).

Because the tectonic processes described above actively shape and reshape the surface, crustal processes produce distinctive geomorphic features that identify landscapes where faulting is (or was) present. However, it is important to recognize the distinction between active and inactive faults, particularly when assessing seismic hazards. An active fault is one along which movement has occurred in historic or recent geologic time. Active faults suggest that movement is likely to reoccur. Inactive faults are typically older geologic features that provide little evidence to suggest that motion has occurred along their traces in historic or recent geologic time. The lack of recent movement gives reason to assume that a recurrence of movement along the inactive fault is unlikely.

The geomorphic features found along fault zones include:

- Fault Valleys
- Saddles
- Scarps
- Linear ridges
- Offset streams
- Sag ponds
- Landslide scars.

Careful interpretation of these landscape features provides useful insight as to where seismic activity occurs and the extent to which fault processes are active in the region. The next step to forming a complete understanding of the hazard involves consideration of its effects. An earthquake can affect the surface directly or indirectly. Direct effects include ground shaking, surface faulting, and displacement. Indirect effects involve processes characterizing one or more forms of ground failure.

Ground shaking is a term used to describe the vibration of the surface and subsurface during an earthquake. The severity of ground motion at a given location depends upon several factors: (1) the total energy released in the form of seismic waves, (2) the distance from the source of the earthquake (epicenter), and (3) the composition of the surface and subsurface geology. This ground motion is characterized by three types of elastic wave (Table 6.5). Typically, horizontal ground movement produces the greatest structural damage during an earthquake. The magnitude and intensity of ground shaking is measured in either of

Table 6.5 Types of seismic wave.

P wave	Primary, longitudinal, irrotational, push, pressure, dilatational, compressional, or push–pull wave. P waves are the fastest body waves and arrive at stations before the S waves, or secondary waves. The waves carry energy through the Earth as longitudinal waves, moving particles in the same line as the direction of the wave. P waves can travel through all layers of the Earth. P waves are generally felt by humans as a bang or thump.
S wave	Shear, secondary, rotational, tangential, equivoluminal, distortional, transverse, or shake wave. These waves carry energy through the Earth in very complex patterns of transverse (crosswise) waves. These waves move more slowly than P waves, but in an earthquake they are usually bigger. S waves cannot travel through the outer core because these waves cannot exist in fluids, such as air, water, or molten rock.
Lg wave	A surface wave which travels through the continental crust.

two ways. One approach to determine magnitude uses the Richter scale. The Richter scale measures the vibrational energy of the seismic wave. It is based on a logarithmic scale, and each time magnitude is raised one unit, the amplitude of the seismic wave increases tenfold (Smith, 1992). The intensity of ground shaking is explained according to the Modified Mercalli Intensity scale. Because ground shaking displays a close relationship to structural damage, the Mercalli scale provides a subjective method to categorize intensity based on the extent of physical damage observed (Table 6.6).

Another direct effect of an earthquake is surface faulting or displacement. It has been observed that during larger earthquakes, fault slippage may extend to the surface, resulting in abrupt ground displacement. This displacement along the fault plane may be either horizontal or vertical. In addition to sudden surface slippage, more gradual forms of slippage may be noted. This slower, less pronounced displacement is known as fault creep, and in some cases it may be more significant a hazard than surface faulting since it needn't be accompanied by an earthquake. Fault creep describes the overall motion along an

Table 6.6 The Modified Mercalli Intensity scale.

Scale/level	Observed effect
I	Not felt except by a very few under especially favorable conditions.
II	Felt only by a few persons at rest, especially on upper floors of buildings.
III	Felt quite noticeably by persons indoors, especially on upper floors of buildings. Many people do not recognize it as an earthquake. Standing cars may rock slightly. Vibration similar to the passing of a truck. Duration estimated.
IV	Felt indoors by many, outdoors by few during the day. At night, some awakened. Dishes, windows, doors disturbed; walls make cracking sound. Sensations like heavy truck striking building. Standing cars rocked noticeably.
V	Felt by nearly everyone; many awakened. Some dishes, windows broken. Unstable objects overturned. Pendulum clocks may stop.
VI	Felt by all, many frightened. Some heavy furniture moved; a few instances of fallen plaster. Damage slight.
VII	Damage negligible in buildings of good design and construction; slight to moderate in well-built ordinary structures; considerable damage in poorly built or badly designed structures; some chimneys broken.
VIII	Damage slight in specially designed structures; considerable damage in ordinary substantial buildings with partial collapse. Damage great in poorly built structures. Fall of chimneys, factory stacks, columns, monuments, walls. Heavy furniture overturned.
IX	Damage considerable in specially designed structures; well-designed frame structures thrown out of plumb. Damage great in substantial buildings, with partial collapse. Buildings shifted off foundations.
X	Some well-built wooden structures destroyed; most masonry and frame structures destroyed with foundations. Rails bent.
XI	Few, if any (masonry) structures remain standing. Bridges destroyed. Rails bent greatly.
XII	Damage total. Lines of sight and level are distorted. Objects thrown into the air.

Abridged from *The Severity of an Earthquake*, US Geological Survey General Interest Publication, US Government Printing Office: 1989-288-913.

active fault and is typically characterized by lateral displacement of the surface measured in terms of millimeters per year. Failure to recognize the presence of creep along a fault zone will lead to the gradual deformation of foundations and structures, sidewalks, roads, and underground utility networks.

The indirect effects of earthquakes are most pronounced in geologic settings that are unstable. The strong ground motion generated during an earthquake produces rapid changes in the condition of these unstable materials. The changes produced by ground motion, such as liquefaction and loss of strength in fine-grained materials, contribute to landslides, differential settlement, subsidence, ground cracking, and other alterations at the surface (Griggs & Gilchrist, 1977).

Coastal processes The attractiveness of coastal landscapes has encouraged the widespread development of shoreline and near-shore areas for residential and commercial activities. Coastal environments, however, are extremely dynamic, with active processes continuously changing nearly every facet of the coastal zone. The inherent instability of coastal areas places important emphasis on the processes operating there and how these processes interact with increasing population concentrations within coastal regions. As the coastal landscape is subject to more intensive forms of use, the environmental stress already evidenced in these regions is likely to intensify as well. Therefore, continued use of the coastal environment must account for its dynamic nature and the diverse forces and processes that interact to perpetuate this landscape and maintain its constant state of flux. Two of the more critical active forces in this landscape, particularly when viewed with respect to hazard potential, are beach-forming processes and coastal erosion (Klee, 1999).

Beach processes and beach formation

The shoreline undergoes change as a result of seasonal and storm induced cycles. The beach, therefore, is a buffer zone that shields coastal areas from the direct forces of wave action. The sources and losses of beach sand, the interaction of waves, tidal action, and wind help determine the size, shape, and extent of beach and the changes these features will experience. Three general conditions can be used to describe the prevailing state along the shoreline (Griggs & Gilchrist, 1977):

1 Accretion predominates over erosion as the beach prograges and builds seaward.
2 The shoreline is stable and neither erosion nor accretion dominates.
3 Erosion predominates and losses of beach sand exceed supply.

Several factors are important regulators of the formation of beaches and greatly influence the rate at which sand is supplied to support beach building. These include stream run-off, sea cliff erosion, and drifting sands from inner continental shelves.

Coastal erosion

The greatest concerns to coastal development are the processes that contribute to coastal retreat and the erosion of coastal areas. Marine erosion is similar in many ways to erosion produced by streams. The most pronounced effects are recognized during short periods separated by longer time intervals where only slight erosion may result. There are four natural mechanisms that direct coastal retreat and erosion:

1 Hydraulic impact – describing the action of waves striking against a sea cliff. This process becomes significant where rocks are well bedded, jointed, or fractured.
2 Abrasion – characterizing the grinding force of beach materials as they encounter coastal rocks.
3 Solution – explaining the wetting and drying of rocks within the intertidal zone. In areas dominated by sedimentary rock, prolonged soaking may contribute to dissolution, hydration, ion exchange, or swelling of grains, loosening them and allowing the material to be washed away.
4 Biological activity – describing the direct mechanical boring, scraping, and indirect

chemical solution of pholads, limpets, and sea urchins on rock material. In some instances between 25 and 50% of the surface area of rock can be riddled with boring, which makes it more susceptible to hydraulic and mechanical erosion.

Coastal erosion becomes a major concern in areas where roads, homes, and other structures have been constructed in proximity to beaches or sea cliffs. In these areas oceanic conditions such as exposure to the sea, wave energy, and the presence or absence of a protective beach play a major role in defining the hazard. Of equal importance is the rock material that comprises the sea cliff or shoreline environment. Here, the controlling factors are rock hardness, bedding, and the density and orientation of jointing. Added to these natural forces are the erosional effects induced by human activity. Human action may accelerate or reduce natural rates in a number of ways, several of the more significant including:

- Loading at the edge of a sea cliff beyond the bearing capacity of the formation.
- Alteration of normal drainage.
- Vegetation and landscaping that promotes physical weathering in rock joints.
- Vehicular vibrations transmitted into a sea cliff from roads or parking facilities.

Volcanic activity Recent evidence suggests that there are approximately 500 active volcanoes in the world. However, as Smith (1992) notes, this figure must be used only as an approximation, simply because it is difficult to accurately determine when a volcano no longer poses a threat. Therefore, as a general rule, any volcano that has erupted within the last 25,000 years can be considered active. With respect to their geographical distribution, the location and behavior of volcanoes is strongly controlled by plate tectonics, with 80% of the world's active volcanoes found in subduction zones where one plate is actively being consumed by another. Along these subduction zones the most prevalent form of subduction volcano is the stratocone volcano. This geomorphic feature produces explosive conditions and reflects the "classic" conceptual image of volcanic activity.

There are, however, many different forms of volcano, each with distinctive eruption patterns (Ritter, 1986).

In considering the nature of volcanic eruptions, magnitude is often measured in terms of explosive capacity, where the release of energy is equated with "*x*" tons of TNT. One method that attempts to quantify this relationship is the Volcanic Eruption Index (VEI). The VEI defines 8 classes of eruption on a scale from 1 to 8. The scale in based on measures of ejecta, height of the cloud column, and related criteria (Nuhfer et al., 1993). In many cases the composition of lava provides a reasonable indication of the explosive potential of a volcano. For example, basaltic magma, because it is low in silica content, high in iron and magnesium and fluid, tends to have a low explosive potential. Rhyolitic magma, however, because of its relatively high levels of silica, acidic nature, and viscosity, has a high explosive potential. Temperature is another factor to consider, particularly with respect to lava flow. When lava reaches the surface at a temperature between 800 and 1,200°C, it will move rapidly, slowing as it cools.

Eruptions, lava flow, and related seismic activity can produce a series of events that pose a hazard. The primary effects of volcanic activity include:

- Pyroclastic Flow
- Air fall Tephra
- Lava Flow.

While secondary effects describe events such as:

- Lahars
- Landslide
- Volcanic gases
- Tsunami.

Hazard, risk, and uncertainty

The natural events discussed in the previous section concern us when they result in negative consequences that harm people or property. Yet the nature of hazardous events, their spatial and temporal distribution, and the exact nature of their consequences and impact are shrouded in an envelope of uncertainty. Put simply, we know that San Francisco is located on the San Andreas fault, but we are not absolutely certain when the next major earthquake will occur, where on the fault it will be located, what its magnitude will be, or what its consequences will be with respect to loss of life or property. To a large degree those answers are subject to the laws of probability and underscore the uncertainty inherent to hazard assessment and planning. For this reason, planning with hazards should operate on the premise that hazardous events are a given; and the planner must work toward a better understanding of risk and develop strategies to narrow uncertainty down to a level that can be more effectively managed.

As Tobin and Montz (1997) note, there is a tendency to erroneously equate risk with hazard. While the concept of risk is an integral part of hazard, the terms are not synonymous. We can define hazard simply as an event that can produce harm, while risk explains the probability that an event will occur. Therefore, risk can be considered as the product of the probability of occurrence and anticipated loss given the hazard. This relationship may be expressed simply as:

Risk = probability of occurrence × vulnerability.

Although this is a useful representation, the formula does not take into account geographic differences in population size or other factors that affect and refine the description of risk. From a planner's perspective, risk is a multiplicative function of hazard, exposure, vulnerability, and response that can be expressed more appropriately as:

$$\text{Risk} = f(\text{Hazard} \times \text{Exposure} \times \text{Vulnerability} \times \text{Response}),$$

where:
Hazard = the occurrence of an adverse event.
Exposure = the size and characteristics of the affected population.
Vulnerability = the potential for loss.
Response = the extent to which mitigation measures are available.

In actuality, risk is more complex than even this functional relationship and extends well beyond the bounded rationality of probability theory. Because the nature of risk associated with a natural hazard can shape both individual and societal perceptions and actions, it is critical to not only understand risk in technical terms, but also to describe how risk is perceived and managed. This more detailed conceptualization places emphasis on the process of risk characterization and its role in hazard decision-making and planning.

Risk characterization

The process of risk characterization has been discussed in detail by Stern and Fineberg (1996). Typically, risk characterization is defined as the process of estimating the consequence of human exposure to a hazard. As a process, risk characterization has traditionally followed as the final step in procedures aimed at assessing risk. Recently, it has been suggested that viewing characterization as a summary stage in assessment carries serious deficiencies (Stern & Fineberg, 1996). This has led to a reformulation of the concept, which now places risk characterization at the very beginning of an assessment process that must:

1 Be decision driven.
2 Recognize all significant concerns.
3 Reflect both analysis and deliberation.
4 Be appropriate to the decision.

In this context, the purpose of risk characterization is to enhance practical understanding and to illuminate choices as they apply to hazards. The ultimate goal of this process is to describe a potentially hazardous situation in as accurate, complete, and decision-relevant a manner as possible, addressing the significant concerns of the interested and affected parties and making this information understandable and accessible to all concerned (Stern & Fineberg, 1996).

Risk characterization begins with the formulation of a problem (the likelihood of harm) and ends with a decision. Overall, characterization entails a five-phase sequence:

1 A judgment sequence – that begins with problem formulation, the selection of options and outcomes, and moves through information-gathering and synthesis.
2 A deliberation sequence – where the limitations and challenges surrounding the hazard

are explored and a set of standards and goals is established.

3 An analysis sequence – where data are assembled and examined to form an understanding of the probabilities associated with the hazard.

4 An integration sequence – that links goals and standards to the characterization of the hazard and its associated risk.

5 An implementation sequence – where policy recommendations and actions are put into motion to reduce vulnerability.

Although a systematic characterization of risk is critical for effective planning, the results of a risk characterization can be misleading if the uncertainty surrounding the process is not interpreted correctly. A common problem when attempting to explain risk relates to the observation that risk characterization often gives the impression of greater scientific certainty or unanimity than truly exists. In these instances, risk characterization may suggest that uncertainty is exclusively a matter of measurement, when in fact its presence may be a matter of disagreement about whether a particular theory applies, or reflect differences in judgment regarding how to infer something that is unknown from something that is known, or simply the impression that certain risks do not exist when in actuality they have not been analyzed (Stern & Fineberg, 1996). The simple truth, however, is that uncertainty pervades all consideration, evaluation, and analysis of risk. Thus, while we strive to eliminate it from the assessment problem, the best we can achieve is a narrowing down of uncertainty until it can be managed.

As an active element of risk assessment, uncertainty derives from the probabilistic nature of occurrences and outcomes and the efficacy of various choices (Tobin & Montz, 1997). Because it is an active element of assessment, it requires special consideration in hazard planning for several reasons:

- It is found in all elements of risk.
- The level of uncertainty is not the same for each element or all hazards.
- Individuals differ in their interpretation and tolerance of uncertainty.
- Reducing uncertainty is not a simple matter.

- Uncertainty may be increased by combined risk.

It is not surprising that the manner by which risk is defined, estimated, and communicated relates to uncertainty. For the planner, understanding this fundamental concept is critical to the formulation of an effective response to hazard.

Uncertainty defines a characteristic presence that surrounds the likelihood, magnitude, distribution, and implications of risk. As a feature of process, uncertainty may arise from a range of factors and conditions. Several of the more obvious sources include:

1 Random variations and chance outcomes of the physical world.

2 Lack of knowledge about the world.

3 An incomplete understanding of process.

4 Lack of an appropriate model of a risk-generating process.

5 Simple ignorance.

A useful taxonomy of uncertainty was presented by Suter et al. (1987). Their taxonomy greatly illuminates the sources and issues surrounding the subject and frames the topic of uncertainty in a manner that can be easily understood. Although it was written with reference to the problem of environmental impact analysis, its concepts are germane to our discussion and can be easily applied to the question of hazard planning and assessment. According to their characterization, uncertainty may be divided into two main classes: defined and undefined uncertainty (Fig. 6.6). Defined uncertainty explains uncertainty intrinsic to the event in question, such as a river flood of a given stage. Undefined uncertainty describes the inherently unknowable and remains largely beyond our consideration. While there is little that can be done to reduce undefined uncertainty, defined uncertainty can be reduced further into two main subclasses, referred to a identity and analytical uncertainty (Suter et al., 1987).

With reference to environmental risk, identity uncertainty defines that uncertainty surrounding the identity of features or individuals affected by an event at some future point in time. Analytical uncertainty explains the uncertainty that develops from the various attempts and methods used to quantify risk and predict events and their

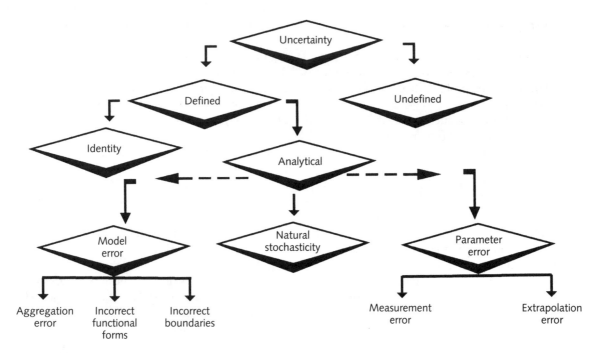

Fig. 6.6 A typology of uncertainty.

consequences. Three important sources of uncertainty can be noted in this context:

1 Uncertainty resulting from the approach used to conceptualize events and their causal mechanisms.

2 Uncertainty that manifests from the natural stochasticity intrinsic to natural processes.

3 Uncertainty associated with measurement error and problems related to the quantification of risk.

To a large degree, risk is an outgrowth of uncertainty. Its presence underscores the need to understand where risk is prevalent and assess risk in relation to the trifold influence of process, consequence, and uncertainty. This is not a simple task given the observation that society all too often overreacts to some risks while virtually ignoring others (Zechhaser & Viscuss, 1990). As a consequence, this pattern suggests that when considering natural hazards, too much importance may be placed on risks of low probability but high salience, risks of commission rather than omis-

sion, and risks whose magnitudes are difficult to estimate (Zechhaser & Viscuss, 1990). Therefore, when one is defining risk in relation to natural processes several important considerations must be incorporated into the risk characterization processes (Whyte & Burton, 1980):

• Risks involve a complex series of cause-and-effect relationships that tend to be connected from source to effect by pathways that may include environmental, technological, and social variables that need to be modeled and understood in context.

• Risks are connected to each other, suggesting that several risks occur simultaneously within the same geographical area.

• Risks are connected to social benefits such that a reduction in one risk usually means a decline in the social benefits to be derived from accepting the risk.

• Risks are not always easy to identify, suggesting that identification frequently occurs long after serious adverse consequences have been experienced.

- Risks can never be measured precisely owing to their probabilistic nature; therefore quantification is always a question of estimation to some degree.
- Risks are evaluated differently over space and time and between various social settings; thus a risk considered serious in one location may be considered unimportant elsewhere.

These factors help to frame risk characterization and direct the methodologies used to conduct an analysis of hazard.

Hazard analysis and assessment

The problem of hazard analysis and risk assessment can be approached in a number of ways. Because the physical hazard and the land use near the hazard may change significantly over time, it is generally necessary to evaluate both existing and future socioeconomic and geophysical conditions within the hazard area. With this information, a clearer determination of the types and magnitudes of damage that may be anticipated from an event now and in the near future can be described.

Practical limitations aside, an appropriate evaluation of a hazard requires a determination of the probability of occurrence of a natural event that poses a threat together with its various intensity levels (Petak & Atkisson, 1982). Here, intensity relates to parameters such as wind velocities, water depth, ground shaking, and slope movement. Because natural events are always active, only those intensity levels capable of producing significant damage are of concern. In this context, hazard intensity is related to the integrity of exposed structures where vulnerability to a given intensity level becomes a function of the design, material composition, occupancy, and construction and maintenance practices of structures in the region.

To effectively assess the levels of risk associated with exposure to a natural hazard, analysis must be conducted in such a manner to provide information on 6 critical elements of risk (Petak & Atkisson, 1982):

1 Identification and description of the characteristics, geographic distribution, potential effects of hazardous events common to the planning area.
2 Assessment of the vulnerability of several classes of building and their occupants to each hazard identified.
3 Identification and measurement of the major primary, secondary, and high-order effects associated with exposure by geographic location of buildings and their occupants to each hazard event.
4 Identification and explication of the major candidate public problems associated with the effects identified.
5 Explanation of the costs and characteristics of the strategies available for mitigating effects induced by exposure.
6 Description of the public policy instruments that may facilitate hazard mitigation.

A comprehensive procedure for conducting a hazard assessment consists of a sequence of analytic stages that expand on the 6 elements listed above (Fig. 6.7). The major steps suggested by this sequence include:

- Hazard analysis
- Vulnerability analysis
- Loss analysis
- Risk analysis.

Although risk assessment may be approached following a clearly defined systematic methodology, a complicating factor throughout analysis stems from the contrasting meanings attached to the concept of risk. These differing interpretations greatly influence what it is that is being measured and how. Therefore, risk may be expressed in an assessment in one of four ways (Tobin & Montz, 1997):

1 Real
2 Statistical
3 Predicted
4 Perceived.

Regardless of definition, the goal of analysis is to define management strategies to minimize, distribute, or share the potentially adverse consequences of a hazard, and suggest options to manage risk in the future. Here, the integration of hazard assessment and mitigation with the local

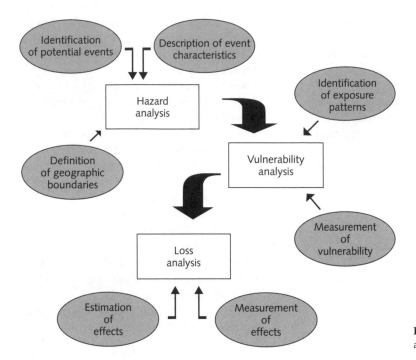

Fig. 6.7 The comprehensive risk assessment process.

land-use and environmental planning becomes critical.

Planning with hazards and risk

Hazard mitigation and environmental planning share several themes in common (Godschalk et al., 1998): both are future oriented, both are concerned with anticipating tomorrow's needs, both are proactive, both attempt to gear immediate activities to longer-term goals and objectives. Traditionally, hazard mitigation, however, has been achieved by either: (1) adopting a locational approach to land use that reduces future losses by limiting development in hazardous areas, or (2) adopting a design approach which focuses on promoting safe construction practices in hazardous areas. The logic behind these mitigation strategies is relatively straightforward. By managing the location of land use, local governments work to shift existing development away from zones of known hazard and direct new land uses toward areas that

are hazard free. Through design management, communities can employ a mix of regulatory measures such as building codes or special-purpose ordinances to reduce damage or draw on nonregulatory programs to inform and instruct developers on the use of damage-reducing design techniques. With most hazard mitigation measures, land-use management has had to contend with a range of problems that have limited its use and reduced its effectiveness (Burby, 1998). These include:

- Problems in maintaining commitment.
- Shortfalls in management capacity.
- Lack of private-sector compliance.
- Failure to act at a regional scale.

As a consequence, these barriers suggest that land-use management alone may not be sufficient to adequately address natural hazard mitigation. Instead, focus has shifted toward a local planning approach that strives to create the appropriate combination of control measures with an emphasis on their effectiveness, efficiency, equality, and feasibility (Burby, 1998). This planning for mitigation approach reflects a significant departure from the practice of hazard management in several important ways. Chief among these is the recognition that appropriate mitigation will vary depending on local circumstance. Therefore, prescription from federal or state agencies down to the local level may overlook critical factors and not reduce vulnerability as intended. According to Burby (1998), taking a local planning approach is one means to ensure that:

- Information on the nature of possible future hazard events is available to the public.
- Land subject to natural hazards is identified and managed in a manner compatible with the type, assessed frequency, and damage potential of the hazard.
- Land subject to hazards is managed with due regard for the social, economic, aesthetic, and ecological costs and benefits to all stakeholders.
- All reasonable measures are taken to avoid hazards and potential damage to existing properties at risk.
- All reasonable measures are taken to alleviate the hazard and damage potential resulting from development in hazardous areas.

Mitigation planning, therefore, combines technical analysis and community participation to enable communities to choose between alternative strategies for managing change. In this context the planning program is intended to produce a plan for avoiding or mitigating harm from natural events and for recovering from their consequences (Godschalk et al., 1998). The general design of this approach can be summarized as a series of choices that direct attention toward:

1. The approach taken to encourage stakeholder participation. This initial step in hazard mitigation planning involves enlisting community support and assistance in formulating the plan. Because some mitigation measures may be controversial, building public involvement promotes awareness. Awareness can include media campaigns, public-school information kits, or homeowner/developer seminars aimed at informing and motivating the community to address natural hazards. As awareness of the hazards improves, more collaborative involvement of the community can be realized. This helps establish meaningful mitigation goals and objectives by combining technical planning, public participation activities, and political action together with a focus on implementation.

2. The emphasis given within each component of the plan. An effective hazard mitigation plan will consist of four essential elements: (1) an intelligence component that will define the problem and provide justification for the policies and actions recommended in the plan, (2) a goals component that consists of a statement of community values as a basis for the policies and actions recommended in the plan, (3) an actions component that provides the details of the policies or programs of action designed to achieve the desired mitigation goal, and (4) an evaluation component that explains how hazard assessment and implementation of the recommended mitigation strategies will be monitored and evaluated.

3. The type of plan. Because plans vary in styles, formats, and emphases, not all types of plans are equally suitable for hazard mitigation. The choice of plan involves selecting an approach to fit the preferences of the community. There are two alternatives to select from: (1) developing the mitigation plan as a separate, stand-alone document focusing on hazards, or (2) incorporating hazard mitigation into the comprehensive community plan. If integration into the comprehensive plan is selected, then the next choice to be made determines whether the plan will be structured as a land classification plan, a future land-use design, a verbal policy plan, a land-use management plan, or a hybrid of the above types.

4. The mitigation strategy. The final area of choice in developing the hazard plan involves selection of the mitigation strategy that will be applied by the community. This aspect of hazard planning is far more substantive in nature and directs attention toward the selection of a particular strategy to achieve mitigation of the hazard(s).

Choice involves consideration of the pragmatic (Godschalk et al., 1998):

- Taking a coercive approach versus a cooperative approach to influence private-sector behavior.
- Employing one local governmental power over another.
- Shaping future development versus addressing existing developments at risk.
- Controlling the hazard versus controlling human behavior.
- Taking action before a disaster occurs versus taking action afterward during recovery.
- Going it alone versus taking an intergovernmental and regional approach.

Although mitigation is often considered synonymous with increasing structural integrity, it is far more useful to view mitigation as a management strategy that balances current actions and expenditures with potential losses from future hazard occurrences. Thus mitigation activities identify those approaches that either (1) eliminate or reduce the probability of occurrence of a hazardous event, or (2) reduce the impact of the hazard occurrence. Two broad categories of mitigation can be described:

- Preparedness activities – define the role of government, organizations, and individuals take in developing, testing, and maintaining programs to save lives and minimize disaster damage.
- Response activities – explain programs designed to provide emergency assistance following a disaster and reduce the probability of secondary damage.

A useful taxonomy of mitigation measures was offered by Petak and Atkisson (1982). According to this schema, mitigation can be divided into seven categories:

1 Approaches that involve measures to minimize the probability of hazard occurrence and/or to protect areas and building sites from the hazard.
2 Approaches that focus on strengthening buildings exposed to the hazard or on the design of site-level systems for protecting buildings from hazards.
3 Approaches that give attention to site development schemes for protecting structures from hazards.
4 Approaches that involve the identification of hazard-prone sites and the application of measures to prevent or restrict their development and use.
5 Loss recovery, relief, and community rehabilitation approaches.
6 Hazard warning and population evacuation systems.
7 Approaches that provide policy-makers with the information and decision assistance tools that facilitate rational and effective hazard management decision-making.

With a focus on mitigation, coordinated planning with hazard suggests a departure from simple hazard management to a more comprehensive view of the problem that encourages a change in the accepted norms of a society (Tobin & Montz, 1997). This new management model, illustrated in Fig. 6.8, suggests that through the combined influence of planning, changes in perception, and the sociological and economic realities faced by communities, new concepts of risk and vulnerability will emerge. For the planner, hazard management is more than ameliorating geophysical events, but also requires modifying the human use system. To this end Blaike et al. (1994) offer 12 principles that can be used to guide hazard planning efforts:

1 Vigorously manage mitigation.
2 Integrate elements of mitigation into development planning.
3 Capitalize on a disaster to initiate or develop mitigation.
4 Monitor and modify to suit new conditions.
5 Focus attention on protection of the most vulnerable.
6 Focus on the protection of loves and livelihoods of the vulnerable.
7 Focus on active rather than passive approaches.
8 Focus on protecting priority sectors.
9 Emphasize measures that are sustainable over time.
10 Assimilate mitigation into normal practices.

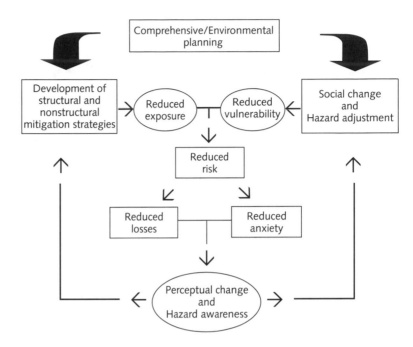

Fig. 6.8 The planning with hazards model.

11 Incorporate mitigation into specific development projects.

12 Maintain political commitment.

Summary

Natural processes often present risks and hazards to humans and should influence the form and location of development. As important responsibility of the environmental planner involves identifying the type and incidence of natural hazards common to the planning area and communicating the risks associated with each to decision-makers. A selection of natural hazards were reviewed in this chapter with a focus on their origins and consequences. From this discussion the concept of risk was introduced and the uncertainties inherent to risk assessment methods were examined. The chapter concluded with a treatment of the planning approaches available to better manage risk and reduce the adverse effects of hazardous events.

Focusing questions

How is risk different from a hazard?

Events that represent hazards have important characteristics. What general characteristics do hazards share?

Which hazards presented in this chapter are common to your planning area and what methods have been used to communicate the risk associated with each?

What does it mean to plan with hazards and how does this depart from traditional planning approaches?

Discuss the role of uncertainty in risk and hazard management: what effect does it have on policy-making?

CHAPTER 7

Environmental Modeling and Simulation

The use of models in planning has enjoyed a long tradition dating back well over four decades. Interest in the use of models is based largely on two familiar planning themes: (1) the need to consider the implications and consequences of decisions before actions are taken, and (2) the requirement to incorporate a large number of interacting variables into a process view of the planning area. Added to these general rationales for using models in planning is the recognition that environmental and regional planning have undergone significant changes that have had a profound impact on the value models enjoy as decision-support aids (Klosterman, 1998). One noteworthy change has been the widespread adoption of low-cost tools and data, such as geographic information systems. However, as these systems became more accessible, it soon became evident that an information system alone cannot meet all the needs of the planner (Harris & Batty, 1993). This realization has renewed interest in the art and science of computer modeling and has stimulated development of integrated tools to support planning and environmental decision-making (Holmberg, 1994; Lein, 1997). In this chapter the fundamental principles of modeling and simulation are introduced, and the role of models in environmental planning is examined with emphasis given to the issues of model design and application.

Models and modeling

The activity of modeling and the creation of models are neither new nor necessarily complex. Since the very early days of our existence, humankind has been urged to interact with our environment to fulfill essential needs. Through this interaction we developed ideas and concepts as to how our environment worked and how we could navigate our world to recover the resources we required to ensure our survival. Over time, these "models" of interaction have become more varied and sophisticated to define more "mature" forms of interaction with the "real world" (Spriet & Vansteenkiste, 1982). Within the realm of science, the interaction between humans and our environments can be approached using either formal or abstract representations. Such representations define models. However, models are only useful if they capture the essential features of the objects, events, or entities they have been designed to represent.

As a device for aiding human insight and comprehension, modeling describes a process that consists of two basic steps: (1) model-building and formalization and (2) model analysis and application. A simple representation of this relationship is given in Fig. 7.1. As suggested by Fig. 7.1, all models begin with the real world. This real world is then simplified to form a representation of key processes that influence the behavior of something we wish to learn more about (climate, land-use change, population growth, ecosystem functioning). Our representation may take several forms, yet regardless of form, a model is produced that stands as a formal characterization of something we are keenly interested in learning more about. This characterization is then examined with reference to the real-world system on which it is based and, if accepted, the model can be used to help understand how that real-world system behaves. The model produced through this process is simply an abstraction of the object, system, or idea in some form other than that of the entity itself, and has value because it can help us to explain, understand, or improve some facet of the real system it represents.

When placed into the context of environmental planning, a model may explain an exact replica of the object (system) at a reduced scale, or it may represent an abstraction of the object's salient properties. In either case models serve several important functions. For the planner perhaps the most useful are prediction and comparison, although there are many other equally relevant functions models fulfill:

- An aid to thought
- An aid to communication
- An aid to experimentation
- An aid to training and instruction.

Model-building, therefore, provides a systematic, explicit, and efficient way for experts and decision-makers to focus their judgment and intuition in a highly structured way. Because a model is a simplification of reality, it may be the only way to understand systems whose geographic scale or complexity might otherwise place them beyond our physical or mental grasp (Hardisty, Taylor, & Metcalfe, 1993). For example, if we wanted to understand an air pollution problem it would be infeasible and unpopular to burn large quantities of coal to see how air quality would change. Thus when we are confronted with a complex relationship, the model may be the only solution. Of course the model is a simplification. Therefore, to be credible the model must retain the significant features of the process in question so its behavior closely approximates what we might find in the real system. Using our air pollution analogy from above, this means that our model should incorporate as many of the properties of the atmosphere, its chemistry, and flow characteristics as possible, so that a change in emission levels can be traced through the model to an outcome that can be evaluated. This point has significant implications for those who use and develop models. Since models are approximations, there is a highly selective and subjective aspect to their design, structure, and purpose.

Several classification schemes have been devised to help understand the types of models that

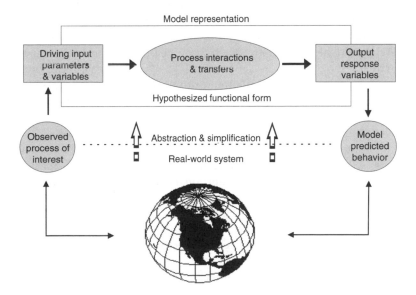

Fig. 7.1 Modeling the "real" world.

have been developed and the nature of the systems they represent. A very simple division separates models into two general groups: descriptive and normative. A descriptive model offers a stylized portrayal of reality with either an emphasis on equilibrium structural features (Static), or on changes in processes or functions over time (Dynamic). Normative models employ the use of analogues. Models may also be conceptual, theoretical, symbolic, mathematical, or statistical. Generally, three basic types of models can be identified (Table 7.1). Regardless of structure, it is the complexity of environmental systems that compels the development and use of models. Furthermore, because environmental planning problems are multivariate and dynamic, relying exclusively on professional judgment does not provide enough support for decision-making. As shown by Vlek and Wagenaar (1979), within any decision problem there is typically a discrepancy between the existing state and the desired state of a given system. While reducing this discrepancy is the goal, there is often more than one possible course of action. Such decision problems are common in planning, particularly when one is considering the environmental consequences of a decision, the pattern of change a decision may induce, or the irreversible commitment of resources that may follow from a decision. In each of these examples the potential number of variables involved is likely to be large, the interrelationships complex, and the uncertainty surrounding the problem high, and the role of models becomes obvious. We would like to explore a "future" before it arrives.

Certain features of a planning problem will point to the appropriateness of a model-based solution. These conditions include:

- A large number of decision, exogenous, and state variables.
- A large number of components.
- Complex and nonlinear functional relationships.
- High degrees of risk and uncertainty.
- A hierarchical structure.
- Multiple and often conflicting objectives.
- Multiple decision-makers.

Applying models in environmental planning, however, is not without its limitations. Critics

Table 7.1 Basic types of models.

Type	Description
Hardware model	Models that take the form of scaled analogue representations of some physical system
Conceptual model	Models that take the form of charts, pictures and diagrams depicting system arrangement and flow
Mathematical model	Models that take the form of numerical expressions that represent critical aspects of process, physical laws, and measured values and relationships
Digital model	Models that describe mathematical models that have been translated into a computer language and encoded for machine execution

of model-based approaches to problem-solving focus on the argument that models are often overly simplistic representations of very complex real systems. Due to their simplicity they are prone to inaccuracies and tend to distort decisions involving important planning resources (Gordon, 1985). Compounding the issue of accuracy is the general observation that far too often the wrong model is selected or the selected model is applied inappropriately. When this happens the results obtained via the model can lead to intractable errors in analysis or inaccurate conclusions. The appropriate use of models in planning depends on whether the model simulates the consequence of realistic decisions, has been validated as an accurate representation of the real world, is reliable, and expresses reliability according to its defined limitations. If these conditions are met, then the model can provide valuable support in making a decision.

When considering the use or development of a model, application typically begins with a problem that involves prediction or comparison as part of the answer. Although there are no formal rules to direct the design of a model, the approach to successful implementation follows the general principle of elaboration and enrichment. According to this simple strategy the model evolves from a simple representation of the processes involved toward a more detailed representation that reflects the complexities of the process more clearly (Lein, 1997; Shannon, 1975). Therefore, modeling involves constant interaction and feedback between the real-world system and its representation. As this model evolves, it is tested, refined, and validated until a useful approximation of the system of interest emerges.

The iterative nature of model development suggests that regardless of sophistication all models are approximations that contain critical assumptions pertaining to the system they represent, the pattern of cause and effect used to capture how the system behaves, and the functional relationships depicted by the elements contained in the model. As an analytic procedure, modeling consists of four key activities (Shannon, 1975): the ability to (1) analyze a problem, (2) abstract from the problem its essential features, (3) select and modify basic assumptions, and (4) enrich and elaborate the initial design. At the completion of these phases, the resulting model can be applied to simulate a real system, and through simulation insight can be gained that helps us understand how the real system responds to change.

The simulation process

An integral feature of the modeling process is the representation of some aspect of the real world using the constructs of systems theory and the placement of the resulting model into an experimental design targeted toward an understanding of that system's future state. These two qualities of modeling connect us back to the systems view of planning introduced in Chapter 1, and move us forward into the realm of computer simulation. Thinking back to Chapter 1, the attraction of systems thinking was its facility for structuring and understanding the behavior of complex interrelationships. When applied to a problem, systems thinking led to the design of a systems model where the variables to include in the system were specified, the hypothetical relationships among variables comprising the system were explained, the structure of the system was described, and the design of a functional form was tested and refined. Using the representation's tools of systems analy-

sis, this initial design can be translated into a "working" model that can be applied toward the solution of a problem.

Designing a model of a real system and conducting experiments using this model describes the general process of simulation. As an experimental and applied methodology, simulation modeling seeks to: (1) describe the behavior of systems, (2) construct theories or hypotheses that account for an observed behavior, (3) use these theories to predict future behavior. Because this is an applied methodology, simulation involves both the construction of the model and the analytical use of the model for studying a problem. Analysis and experimentation are central themes in this process as the problem, filtered through the model, is examined in a controlled and systematic way to reveal new insights and test present assumptions (Fig. 7.2). These central concepts also connect simulation to the goals of planning by providing an analytic device that can test alternatives, evaluate allocation strategies, and examine critical trends within the planning area.

To successfully apply computer simulation methods in environmental planning, three interrelated activities must be understood:

1 Model Design – the initial stage in the simulation process that includes a detailed formulation of the problem, a clear definition of the system, and the specification and testing of a model.

2 Model Application – the second stage in the process that includes calibrating the model, selecting the scenario to examine, and executing the model.

3 Analysis and Implementation – the final stage in the simulation process that concerns the interpretation of the results obtained from the model and the implementation of those results

Given the methodology imposed by these activities, a simulation experiment can be looked at as a procedure for acquiring information. This means that through the experimental run of the model, it will produce results that can provide insight into the problem, and these results can be data in the form of numerical approximations of some change in a variable, or it can be a graphic representation of how something may look given the

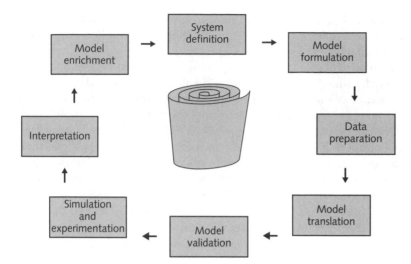

Fig. 7.2 The process of simulation modeling.

inputs that were selected to drive the model. Making certain that this information is useful and its limitations understood introduces several important considerations that guide the simulation experiment:

- Validation and Verification – describes a series of tests applied to the model in order to establish a level of confidence in the model and the inferences drawn from the simulation. Because a model is essentially a theory describing the structure and interrelationships of an observable phenomenon, validation is not concerned with whether the model is a "true" representation of the actual system, but rather whether the insights gained from the simulation experiment are reasonable. Through validation it is possible to explain the operational utility of the model. This may involve several specific tests to build confidence, such as face validity to determine if the model results appear correct and assumption testing to ascertain whether critical assumptions in the model can be supported.
- Sensitivity Analysis – because a model receives input data and processes that data to generate an output, how the model responds to the variables and parameters used in a simulation greatly influences its useful application. As a test, sensitivity analysis

consists of systematically varying the values of the parameters and input variables over a range of known extreme cases. The analyst can then observe the effect of these extreme values on the model performance and the results it provides. Through sensitivity testing: (1) limitations in the model due to the values used to parameterize and conduct a simulation can be identified; (2) effects produced by extreme conditions on the stability of the model can be noted; and (3) clues to guide future enhancements and modifications of the model can be located.
- Experimental Design and Execution – running a simulation experiment is the process of applying the model to observe and analyze the information it provides. The experimental design selects a specific approach for gathering the information needed to enable the environmental planner to draw valid inferences. Three essential steps direct this process (Shannon, 1975):
  1 Determination of the experimental design criteria.
  2 Synthesis of the experimental model.
  3 Comparison of the model to standard experimental designs to select the appropriate methodology.

Typically, the simulation experiment concerns resolving an answer to the question, "How does a change in x affect y?" In an environmental planning context we may ask how a change in land use might affect surface run-off, or how an increase in vehicle trips will affect air quality. Thus, the final design of the simulation tends to be strongly influenced by the criteria deemed pertinent to that question. Among the criteria to consider when finalizing the design are:

1 The number of factors to be varied.
2 The number of values or levels to use for each factor.
3 Whether the various levels are quantitative or qualitative.
4 Whether the levels of the various factors are constant of random.
5 Whether nonlinear effects will be measured.
6 The number of measurements that will be taken for the response variable.
7 How interactions between factors will be measured.
8 What level of precision is required for an effective analysis?

These eight points remind us that a simulation experiment is only as sound as the techniques used in its construction. This suggests that the validity of the results gained through simulation can be affected by the techniques used in data collection and the methods employed when summarizing the data. Therefore, to be useful, simulation must fit within the context of a clearly defined problem. The problem focus takes the form of a forecast, where the planner seeks insight into the future state of the planning area and the forces of change that will drive the planning area toward this new state. Expressing the problem in the form of a forecast directs attention to the question of prediction and its connection to the modeling and simulation process.

Prediction and scenario projection

Modeling and simulation facilitate one of the main goals of planning, that of prediction. Because a plan is essentially an attempt to explain how the landscape will evolve through the implementation of specific goals and policies, the ability to predict what might happen if a clearly defined policy choice is selected serves an essential analytical need. Prediction, of course, does not necessarily imply that future conditions will be forecast exactly. Rather, prediction implies that certain assumptions about the future can be explored and evaluated. Modeling and simulation support this type of evaluation by creating an environment where a range of alternatives can be examined and a range of potential future arrangements of the built environment can be explored.

According to Rescher (1998), any prediction worth considering must rest on an evidential basis. This suggests that some rational substantiation must exist, because serious cognitive interest attaches to rational prediction, and for those that are credible there is good reason to accept them as correct. When predicting, there must be an actual commitment to a future-oriented claim. Consequently, the predictive character of those future-oriented declarations depends on the user of those declarations. However, not every future-oriented thesis represents a prediction. Thus, for a claim to constitute an actual prediction, there must be someone who makes or accepts it. Consequently, prediction differs in its objectives and purpose from scenario projection.

Based on these observations, the concept of a scenario can be defined in several different ways. Koplik et al. (1982) define a scenario as a possible sequence of processes and events that are describable by equations involving specified physical parameters. Other, less technical definitions characterize the concept as

• A hypothetical sequence of events constructed for the purpose of focusing attention on causal processes and decision points.
• An exploration of an alternative future.
• An outline of one conceivable sequence of events and states given certain assumptions.

A scenario, therefore, is a tool for ordering one's perceptions about alternative future environments in which today's decisions might play out. In practice, scenarios resemble a set of stories, written or spoken, built around carefully constructed plots. Stories are an old way of organizing knowledge, and when used as planning tools, they defy denial by encouraging – in fact, requiring – the willing suspension of disbelief. Stories can express multiple perspectives on complex events, and scenarios give meaning to these events. Yet, despite its story-like qualities, scenario planning follows systematic and recognizable phases. The process is highly interactive, intense, and imaginative. It begins by isolating the decision to be made, rigorously challenging the mental maps that shape one's perceptions, and hunting and gathering information, often from unorthodox sources. The next steps are more analytical: identifying the driving forces (social, technological, environmental, economic, and political); the predetermined elements (i.e., what is inevitable, like many demographic factors that are already in the pipeline); and the critical uncertainties (i.e., what is unpredictable or a matter of choice, such as public opinion). These factors are then prioritized according to importance and uncertainty. These exercises culminate in a small set of carefully constructed scenario "plots" (Schwartz, 1991; Mason, 1994; Wack, 1984).

If scenarios are to function as learning tools, the lessons they teach must be based on issues critical to the success of the decision. Since only a few scenarios can be fully developed and remembered, each should represent a plausible alternative future, not a best-case, worst-case, and "most likely" continuum. Once the scenarios have been fleshed out and woven into a narrative, the planner identifies their implications and the leading indicators to be monitored on an ongoing basis. A scenario, therefore, has several distinguishing characteristics that guide its creation and application in modeling. First, and perhaps most importantly, scenarios are hypothetical simply because the future is unknowable. The best one can hope for is to explore alternative possible futures. Therefore a scenario will never materialize exactly as described due to the impact of unforeseen events and responses (Fowles, 1978). Secondly, a scenario professes to be only an outline of a possible future. For that reason, a scenario only seeks to identify the key branching points of a possible future to highlight the major determinants that might cause the future to evolve from one branch rather than another. Thus, the scenario serves to sketch in the primary consequences of a causal chain of events in a highly selective manner (Fowles, 1978). Finally, a scenario should be multifaceted and holistic in its approach to the future.

The test of a good scenario is not whether it portrays the future accurately but whether it enables an organization to learn and adapt. When applied to problems in environmental planning, scenarios are a means to integrate individual analyses of trends and potential events into a holistic picture of a possible future. In this context, scenarios become powerful planning tools precisely because the future is unpredictable. Hence, unlike traditional forecasting methods, scenarios present alternative images instead of extrapolating current trends from the present. Scenarios also embrace qualitative perspectives and the potential for sharp discontinuities that econometric models exclude. Consequently, creating scenarios requires planners to question their broadest assumptions about the way the world works so they can anticipate decisions that might be missed or denied. Thus, the planner creates a scenario to describe the interaction of trends and events and to explore the possible course of alternative decisions on the future state of the planning area. To be effective, a prerequisite for scenario use in planning must be a sensing of incipient societal change, whether those changes are demographic, environmental, economic, technological, or some combination of the above. Given this, the scenario serves as a synoptic view of the total future environmental possibilities and is designed to meet five analytic objectives (Wilson, 1978):

1 To combine alternative environmental development into a framework that is consistent and relevant to the planning area.
2 To identify "branching points," potential discontinuities and contingencies that can serve as valuable early warnings.
3 To formulate strategies that can translate alternative environmental developments into policy recommendations.
4 To provide a basis for analyzing the range of possible outcomes.
5 To test the outcomes of various planning strategies under alternative environmental conditions.

An important consideration when applying scenarios in planning is the fact that there is no inevitability to the future. For this reason, it is critical to present as many varying views of the future as practical in order to define

- Trends that are probable but "shapeable."
- Trends that are probable but not amenable to policy influences.
- Trends that are possible and "shapeable."
- Trends that are possible but not amenable to policy influences.

While there is no single best method to follow when developing a scenario, a prototypical pattern can be offered. In general, developing a scenario requires compiling a detailed listing of all phenomena potentially relevant to the problem. Added to this list are the processes and events that influence change. Armed with this basic information, specific effects and consequences can be identified. Next, the scenario can be explained in written form and reviewed to assess the credibility of the pathways and connections that shape alternative futures. Once expressed in this form, a reasonable number of possible futures can be selected for further analysis.

Because more than one description of the future is possible, a major issue in the use of scenarios in planning relates to the problem of selection. Several methods have been introduced to guide scenario selection: simulation, event-tree analysis, and expert judgment. The relative merits of each have been summarized by Ross (1989). Although specifics differ, selection methods address two distinct functions: (1) ensuring completeness of the list mechanisms that are considered, and (2) screening out irrelevant and incredible mechanisms to produce a list of reasonable scenarios. Of the selection methods introduced, the most useful to the environmental planner is the event tree. An event tree is a scenario constructed in the form of a diagram (Fig. 7.3). These diagrams describe events and processes that explain specific chains of cause–effect–effect–consequence relationships. Using an event tree the logical sequence of cause and effect leading to a specific outcome can be traced. This feature has the advantage of allowing scenario developers to follow the causal sequence and critically review the plausibility of each step, examine the logic and assumptions inherent to an event chain, and evaluate in probabilistic terms the likelihood of the relationship depicted.

Although scenarios are useful tools in planning, they are not without limitations (Godet & Roubelat, 1996). Discussing the use and misuse of scenarios in long range planning, Godet and Roubelat (1996) remind us that a scenario is not a future reality, but a way of foreseeing the future. In a scenario we are using the present to suggest all possible or desirable futures. The future is not being predicted, which implies that our scenarios are only useful if they are relevant, coherent, likely, and transparent. This places tremendous responsibility on the planner, who must ask the

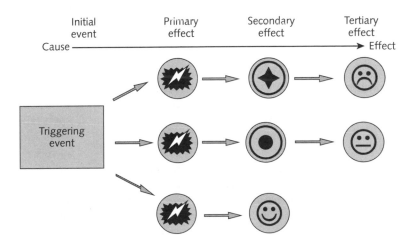

Fig. 7.3 A stylized event tree diagram.

right questions and clearly formulate the hypotheses that are the keys to the future.

Computer modeling methods

Developing scenarios can be a valuable way to gain insight regarding the causal processes that influence the future state of the planning area. However, to make a scenario useful as a predictive device, the facts, decision points, and causal mechanisms must be operationalized to permit the consequence of outcomes to emerge for a set of selected endpoints. Operationalizing a scenario means putting it into motion, letting it run, and observing what its products might look like. Placing a scenario in motion requires developing a simulation experiment to model critical features of the hypothetical sequence of events and using simulation as a means of exploring the output generated by the model. There are various approaches available to model the causal processes implied by a scenario, and numerous methodologies used to capture and characterize causality. Thus, when the focus in planning shifts to the question of modeling, the issue of which method best explains the process of interest must be resolved.

Causal relationships can be defined in either of two contrasting ways: (1) continuous versus discrete processes or (2) stochastic versus deterministic processes. When used to explain process, the terms "continuous" and "discrete" refer to the nature or behavior of the system as evidenced by a change in its state with respect to time. A system (process) whose changes occur continuously over time are termed continuous, whereas systems whose changes in state occur in finite quanta or jumps are referred to as discrete. The terms "stochastic" and "deterministic" refine the nature of how those changes occur. A deterministic process is one in which each new state is completely determined by its previous state of defining conditions. Thus, a deterministic system evolves in a completely fixed and conclusive way from one state to another in response to a given stimulus (Lein, 1997). Stochastic processes contain an element of randomness that influences the transition from state to state. Stochastic systems suggest a behavior that may be probabilistic in nature and explained in terms of a probability function. Based on an understanding of the problem and the causal mechanisms involved, the decision must be made regarding which of the above definitions can be used to represent process in a model. This decision must also take into consideration the geographic scale and the level of spatial resolution needed to adequately represent the process of interest as an active feature of the landscape. Finally, the controlling variables and parameters that drive the system need to be identified, and values for these entities must be obtained and evaluated.

The range of modeling methodologies and their technical specifications are well beyond the scope of this chapter. Excellent treatments of modeling with reference to planning and environmental management can be found in Gordon (1985), Klosterman et al. (1993), and Hardisty et al. (1993). In place of an exhaustive review of modeling methods, this section will examine a selection of modeling "recipes" that can be employed when theorizing about causality and developing the kernel of a simulation study. The selected recipes define fundamental representational schemes for expressing causal processes and approximating the behavior of a system, and include the methods of:

1 Monte Carlo sampling
2 Markov processes
3 Optimization
4 Systems dynamics
5 Cellular automata.

Before embarking on this review, it is helpful to explain the basic process of converting knowledge of a system into a numerical model and to describe how these techniques fit into that process. For the purposes of our discussion, a model is constructed in four phases, each with a specific objective.

1 **Specifying the purpose of the model** – every model is designed to meet a specific need for information. Whether calculating river flow, population changes, or the spatial interaction between land-use zones, execution of the model will give valuable insight into the behavior of the process under investigation.

2 **Specifying the components to be included in the model** – because a model is an abstraction or simplification of reality, the elements that drive the behavior of the system must be identified. These components determine the functional form of the model and explain key interactions that describe process.

3 **Specifying the parameters and variables associated with the components** – the descriptive elements that are used to explain the system must be measurable qualities and quantities for numerical simulation to work. Thus, each element has to be expressed in terms of a value that reflects its present status as well as its dynamic nature.

4 **Specifying the functional relationships among components, parameters, and variables** – this final phase concerns the explicit treatment of process and how the model will capture the behavior of the system. The functional relationships defined here establish flows of information, matter, or energy as directed by the rules used to govern model behavior. The basic functional forms can include any of the following:

a. **Deterministic relationships** – explaining the conditions where behavior is completely determined by the state equations used in the model to calculate the state of selected components.

b. **Probabilistic relationships** – defining the condition where underlying processes expressed in the model can be represented by governing rules of probability.

c. **Stochastic relationships** – characterizing behavior or processes where uncertainty is high and random elements of chance influence behavior. Such relationships are not rigidly controlled by probability, but rather follow a more heuristic pattern with greater potential variability.

The problems encountered in environmental planning typically involve some aspect of all three relationships expressed above. In many situations selecting the correct modeling strategy is not easy and may be influenced by data availability, lack of underlying theory, and the degree of uncertainty inherent to the problem.

To understand how process can be represented and to decipher the "black-box" view of models as analytical tools, their underlying structure and mathematical foundations can be examined. Because many models used in environmental planning share these fundamentals, the techniques demonstrated reveal the elegance and simplicity of the modeling process and illustrate how root ideas can be easily adapted to new situations and problems.

Monte Carlo sampling

Monte Carlo sampling is rooted in the concept of simulating systems containing stochastic or probabilistic elements (Shannon, 1975; Evans & Olson, 1998). Simulation models based in this technique trace their origins to the work of von Neumann and Ulan during the development of the atomic bomb. Although the technique had been known for many years prior to that period, its success at Los Alamos quickly encouraged its application to problems in a range of disciplines. Although the primary use of Monte Carlo sampling is its utility for simulating probabilistic situations, it can also be applied to completely deterministic problems that cannot be solved analytically.

Using the analogy of a Roulette wheel on a gaming table, Monte Carlo sampling employs artificial experience or data generated by the use of a random number generator and the cumulative probability distribution of interest to produce a pattern of numbers that represents the behavior of "real-world" objects, events, or entities. The procedures used to generate random numbers rest at the core of this modeling technique. Typically, random numbers can be acquired from a computer program or subroutine that can provide uniformly random digits. However, in most simulation models we often wish to generate random numbers whose distributions are other than uniform. In these instances the uniformly distributed pseudorandom numbers generated by our program are used to "draw" or sample values from a known frequency (probability) distribution (Harbaugh & Bonham-Carter, 1981). The probability distribution to be sampled can originate from a variety of sources, including:

• Empirical data derived from past records of the event or process (i.e., stream flow, traffic counts, wind speeds).
• A recent experiment or field test that generated measurement values of the event or process.
• A known theoretical probability distribution (i.e., Gausian, gamma, poison).

The random numbers generated based on one of the above methods are used to produce a randomized stream of variates that will duplicate the expected experience (behavior) as a function of the probability distribution being sampled.

Monte Carlo sampling is relatively simple in concept. Shannon (1975) provides an excellent discussion of the major steps involved by means of an example that can be worked by hand (Table 7.2). To draw an artificial sample at random from a population that can be summarized by a probability function, let's consider a hypothetical simulation model dealing with the deposition of solid waste in a municipal landfill. The model in our example is constructed so that the quantity of solid waste deposited per increment of time (years) is an exogenous input to the model. The quantity of solid waste can be defined as the volume (thickness) deposited uniformly over the area of the landfill per increment of time. In designing the model an initial step would be to gather actual data about the frequency distribution of solid

waste quantities at a sample of existing landfills. In practice this might prove to be difficult, but we might have volume data of annual deposition for 500 landfills throughout the United States. The data can be displayed as a series of discrete intervals in a histogram based on the frequency with which measurements fall into each class of the histogram. This empirical distribution can be converted to a cumulative distribution to enable the draw of random variates. Using this cumulative distribution, artificial experience for our landfill can be generated to characterize its life expectancy. Thus to simulate solid waste deposition in a computer model we can sample the cumulative frequency distribution represented by the histogram through the use of a random number generator. This can be accomplished in either of two ways:

1 Random samples can be drawn from the observed empirical distribution itself.
2 If the frequency distribution can be approximated by known theoretical distributions, samples can be drawn from that distribution provided that we can write its function and estimate its parameters (such as the mean, and standard deviation for a normal distribution).

For the present example, let's assume we have drawn our samples from the cumulative distribution. With these samples we would like to evaluate the life expectancy of a candidate landfill site over a 50-year time horizon. Using a random number generator a series of variates can be drawn to represent annual solid waste deposited at the site. If the numbers 09, 57, 43, 61, 20 are drawn for the first five years of operation, then resulting patterns of solid waste can be derived as shown in Table 7.3. Running this hypothetical model for all 50 years in our example, a synthetic process can be generated to simulate landfill performance. The values representing the 50-year sequence can then be used to evaluate landfill capacity. If the random numbers used are in fact uniformly distributed and random, then each number from the data of interest will occur with the same relative frequency as we might expect in the "real world." In this example the artificial experience is typical of what we have been experiencing with the real system.

Table 7.2 Basic steps in Monte Carlo sampling.

Step 1	Plot or tabulate the data of interest as a cumulative probability distribution function with the values of the variate on the x-axis and the probabilities from 0.0 to 1.0 plotted on the y-axis.
Step 2	Choose a random decimal number between 0.0 and 1.0 by means of a random number generator.
Step 3	Project horizontally the point on the y-axis corresponding to the random number until the projection line intersects the cumulative curve.
Step 4	Project down from this point of intersection on the curve to the x-axis.
Step 5	Write down the value of x corresponding to the point of intersection. This value is then taken as the sample value.

Based on Shannon (1975).

Table 7.3 Simulated landfill deposition in millions of tons.

Random number	Simulated range
0.09	0.0–1.00
0.57	4.0–5.0
0.45	2.0–3.0
0.61	4.0–5.0
0.20	1.0–2.0

When one is creating a model that defines stochastic or probabilistic elements, an important consideration in the application of the Monte Carlo method relates to the issue of whether to use empirical data or a theoretical distribution. This question is significant for three reasons (Shannon, 1975). First, the use of empirical data carries the implication that the model is simply simulating the past. Therefore historic data replicates the performance of the system based on patterns that have already occurred. Assuming that the basic form of the distribution will not change over time does not presuppose that the patterns evidenced over a given time period will be repeated. Secondly, it is generally considered more computationally efficient to use a theoretical distribution. Lastly, it is much easier to change the defining parameters of a random number generator based on a theoretical distribution, perform sensitivity tests, and ask "what if" questions. Therefore, a more useful model can be produced using theoretical distributions (Shannon, 1975).

Markov processes

It has been observed that many environmental processes that are random in their occurrence also exhibit an effect in which previous events influence, but do not rigidly control, subsequent events. Such processes are referred to as Markov processes. In simple terms, a Markov process characterizes the condition in which the probability of the process under investigation being in a given state at a particular time can be deduced from knowledge of the immediately preceding state (Harbaugh & Bonham-Carter, 1981). This general characteristic of system behavior is referred to as a Markov chain. A Markov chain can be conceptualized as a sequence or chain of discrete states in time or space where the probability of the transition from one state to a given state in the sequence depends on the previous state. Thus, the general form of a Markov chain explains a series of transitions between different states or conditions of a system such that the probabilities associated with each transition depend only on the immediately preceding state, and not on how the process arrived at that state (Harbaugh & Bonham-Carter, 1981). Based on this premise, a Markov chain will typically contain a finite number of states and the probabilities associated with the transitions between states do not change with time. Consequently, a Markov chain in its general form defines a short "memory" of process that extends only for a single step at a time and stops after that single step. A chain that exhibits this characteristic is termed a first-order Markov chain. This fundamental definition can be extended to describe the condition where the probabilities associated with each transition are based on earlier events of multiple dependence relationships. Perhaps the most important characteristic of Markov chains is that they exhibit a dependence on the probabilities associated with each transition of the immediately preceding state. This quality is called the Markov property, and to apply this modeling technique effectively, the phenomena under investigation should exhibit this fundamental behavior. Fortunately many processes encountered in environmental planning share this trait. Therefore, the planner can make effective use of Markov chains as components in probabilistic dynamic models.

To illustrate how Markov models can be applied in environmental planning we can explore the problem of land-use change and use the Markov property to help forecast the likelihood of future land-use transitions. As demonstrated in previous work, land use and land cover change can be characterized as stochastic (Lein, 1990). This assumption is based on the observation that the physical use of land, while often described using an economic rationale, is ultimately determined by the locational decisions of governments, corporations, and private individuals. Such locational decisions introduce behavioral influences into the land development process, creating the situation where the use of a parcel of land becomes a function of policy decisions, the physical suitability of land, and an intangible set of personal motivations that may or may not be guided by economic incentives. The consequence of this complex series of operations is a pattern of land occupance that when viewed collectively represents a series of random elements acting in space. Thus in a stylized way the pattern of land-use change can be explained as stochastic where past trends can influence the future state of the system (Bourne, 1971; Bell, 1974).

With the process of land-use change now characterized as Markovian, a parcel of land (L) will define a specific condition or form of use. That condition describes a particular state (s_1) at that instant in time (t). Associated with that parcel (L) is a set of probabilities $(p_1, p_2, \ldots, p_n)$ that express the likelihood of parcel (L) occupying the same or a different state $(s_1, s_2, \ldots, s_n)$ at time $(t + 1)$. In this example there is a given set of states, each representing a category of land use, and the implicit requirement that parcel (L) can be in one and only one state at time (t). However, based on the probability of change, parcel (L) can move successively from one state to another. Consequently, the probability of change projects parcel (L)'s position in the system and defines the condition it may assume.

According to the Markov chain model, the set of probabilities characterizing the land-cover system is expressed in the form of a transition probability matrix [Z]. This matrix describes the basic behavior evidenced by the system as expressed by the Markov chain and defines the principal pattern of movement as elements in the system change from state to state. Each element in the matrix, therefore, reflects the probability of a transition from a particular state (that state pertaining to a given row in the matrix) to the next state (that state pertaining to the particular column) (Harbaugh & Bonham-Carter, 1981).

For a simple three-state system the transition probability matrix may be written as:

$$
\mathbf{Z} = \begin{array}{c} \\ s_1 \\ s_2 \\ s_3 \end{array}
\begin{array}{ccc}
S_1 & S_2 & S_3 \\
\left[\begin{array}{ccc}
Z_{11} & Z_{12} & Z_{13} \\
Z_{21} & Z_{22} & Z_{23} \\
Z_{31} & Z_{32} & Z_{33}
\end{array}\right]
\end{array}
$$

where each element in the matrix explains the probability of movement from one state to another. In our land-use example, that might look something like this hypothetical case:

	Urban	Agric.	Forest
Urban	0.9	0.1	0.0
Agriculture	0.6	0.3	0.1
Forest	0.4	0.4	0.2

In this simple example, the matrix suggest that if the land cover system is in state s_1 (urban) its most probable next state is urban (s_1). However, if the system is in state s_2 (agriculture), a transition to urban is probable, as is the transition to forest (0.1). Finally, if the land-use system is in state s_3 (forest), a transition to either urban or agriculture is equally probable. Here we can see that the simulation of process involves use of the transition probability matrix to select the state of the system. In this example, to simulate change in the land-cover system the matrix of transition probabilities must be computed and the resulting values used to produce a representation of the system at time $(t + 1)$.

Transition probabilities are derived from the frequency distribution of the objects that comprise the various states of the system. Producing this distribution of state transitions involves nothing more than the tabulation of the number of transitions (moves) from each state to every other state in the system. Computationally the process involves filling a tally matrix, counting the movement of objects from state to state, then converting that matrix to transition probabilities by dividing by the sum of the rows. The calculation of a transition probability matrix, however, explains only a single step in a Markov chain. For a multiple-step transition defining the transition over more than one time step, the series of probabilities is determined by "powering" the transition probability matrix [Z]. The term "powering" simple means to raise the matrix [Z] to some power. The product of powering the matrix reflects the transition to the next time step in the simulation. The general form of the succession of matrix powering can be written as:

$$Z_{(n)} = Z_{(n-1)} \cdot Z.$$

In the present example we can use a one-step process to produce a distribution of lands uses for $(t+1)$ from the patterns that explain the system at time (t) (the present state of the land-use system) and time $(t-1)$ (the land-use pattern one step back in time). As used in our example, the condition explained above assumes that the process of land-use change is a discrete-time phenomenon. Therefore, while the surface is dynamic, change is observed by using discrete slices of time.

We can extend this idea to long time sequences. For instance, if we chose to examine the case where the system beginning in state (i) would be in state (j) after n steps, the Markov chain could be expressed as:

$$\mathbf{Z}_{(n)} = \begin{bmatrix} Z_{11}^{(n)} & Z_{12}^{(n)} & Z_{13}^{(n)} \\ Z_{21}^{(n)} & Z_{22}^{(n)} & Z_{23}^{(n)} \\ Z_{31}^{(n)} & Z_{32}^{(n)} & Z_{33}^{(n)} \end{bmatrix}$$

The solution to this problem is easily derived by use of matrix algebra. The matrix notation demonstrating the solution to our problem is shown in Table 7.4.

Optimization

Optimization is one of several methodologies that fall within the general subject matter of operations research. Operations research is a name given to the procedures used to study and analyze problems concerned with the control or operation of systems. Within this broad definition, operations research explains the systematic application of quantitative methods, techniques, and tools with the goal of evaluating probable consequences of decision choices. The decisions considered using optimization typically involve the allocation of resources with the objective to improve the effectiveness of the system. Therefore, most operations

Table 7.4 Markov analysis matrix notation.

---
1. Initial step:
   Z
2. Raising to the second power:
   $Z^{(2)} = Z \times Z$
3. Raising to the third power:
   $Z^{(3)} = Z^{(2)} \times Z$

---

research problems center around the optimization of a specific feature or flow in the system, such as goods, information, technical properties, capacities, or any tangible characteristics that lends themselves to strategies where "costs" are to be minimized and benefits maximized given a set of constraints.

The essential characteristics of operations research include:
- Examining function relationships in a system.
- Adopting a planned approach to problem-solving.
- Uncovering new problems for study.

With respect to the problems encountered in environmental planning, operations research can be thought of as a form of applied decision theory where the collection of mathematical techniques and tools is used in conjunction with a systems perspective to address decision problems. Certain attributes of a problem may lend themselves to treatment using optimization, particularly those involving complex relationships that can be abstracted into the form of mathematical or statistical models.

Any operation research application draws on five common phases of analysis:
1. Formulating the problem.
2. Constructing a mathematical model to represent the operation under consideration.
3. Deriving a solution to the model.
4. Testing the model and evaluating the solution.
5. Implementing and maintaining the solution.

Because environmental planning is concerned with the problem of maintaining an optimal balance between social needs and environmental process, methods that can provide insight regarding the "optimal" allocation of resources can greatly assist the decision process. Viewing optimization at its most fundamental level, most methods are applied to human-designed or controlled systems where some optimal situation is the goal. Although determination of what precisely defines the "optimal" relies exclusively on human judgment, the majority of techniques direct efforts toward the maximization of benefits. Analytical optimization techniques, therefore, are applied in a problem-solving context where the need exists to either maximize or minimize a clear and narrowly defined objective (the objective function).

Satisfying this objective function recognizes that any system encountered by the planner operates with constraints that limit options. Such constraints may identify design limitations, carrying-capacity measures, safety considerations, or criteria that influence biological functioning. For example, a planner may desire a subdivision plan that utilizes land as efficiently as possible at the lowest environmental cost. Here, cost can be expressed in relation to habitat fragmentation and "efficiency" becomes the objective function given the constraint that the plan cannot increase fragmentation beyond a threshold. From this simple example it can be seen that the theory of optimization shares several similarities to the concepts embedded in the idea of system control. However, there are significant differences that should be understood. Perhaps the most important distinction is that system control is primarily concerned with the response of a system under conditions of fluctuating inputs. Optimization considers the definition and properties of an objective function that may be applied as the criterion for control. While subtle, this distinction suggest that optimization methods are applied in a more prescriptive manner.

A wide array of optimization methods are available to select from, and most are variations of the mathematical programming model. Mathematical programming is rich in its facility for resolving objective functions. In our discussion here we will limit our scope to those programming methods that are linear in nature. The methods of linear programming form two broad categories: (1) direct search and (2) indirect search. Each of these classes can be further divided into more specific algorithms. Regardless of the algorithm applied, the focus remains fixed on the concept of an objective function. This term defines a dependent variable (Y) whose value depends on one or more independent variables ($x_1, x_2 \ldots , x_n$). The goal of optimization is to either maximize or minimize the objective function by finding the values of the independent variables such that:

$$Y_{(max)} = f(X_1, X_2, \ldots , X_n).$$

Or

$$Y_{(min)} = f(X_1, X_2, \ldots , X_n).$$

For optimization problems where the objective function is a linear combination of the decision (independent) variables and subject to linear inequality constraints, its solution can be by obtaining values of $(x_1, x_2, \ldots , x_n)$, so that the linear function

$$Z = c_1 x_1 + c_2 x_2 + \ldots + c_n x_n$$

is either maximized or minimized subject to the constraints:

$$a_{11} X_1 + a_{12} X_2 + \ldots + a_{1n} X_n <= b_1$$
$$a_{21} X_1 = a_{22} X_2 + \ldots + a_{2n} X_n <= b_2$$

.

.

.

$$a_{m1} X_1 + a_{m2} X_2 + \ldots + a_{mn} X_n <= b_m$$

where:

$$x_1 >= 0, x_2 >= 0, \ldots , x_n >= 0$$

and a_{ij}, b_i are given as constants.

One widely used approach for solving linear programming problems is the application of the Simplex method (Hillier & Lieberman, 1967). An excellent example illustrating the basic principles of the Simplex method can be found in Harbaugh and Bonham-Carter (1981). According to this technique, given two decision variables, a relation can be expressed where

$$Z = 4x_1 + 3x_2 = 24,$$

and the variable Z is to be maximized subject to the constraints that

$$x_1 \leq 5, \text{and } x_2 \leq 4$$

and

$$4x_1 + 3x_2 \leq 24 : x_1 \geq \phi ; x_2 \geq \phi.$$

Representing the problem graphically, the constraints x_1 plotted against x_2 for an area that defines the feasible solution space (the area shaded on Fig. 7.4). By plotting the values of the objective function across the solution space the optimum solution can be defined as the point (x_1, x_2) at which Z is maximized and falls within the area of the feasible solution. Careful interpretation of the figure identifies the point as the apex of the shaded region where $x_1 = 3$ and $x_2 = 4$.

Systems dynamics

Systems dynamics, or its more recent incarnation as dynamic modeling, refers to a family of models and an integrated modeling environment designed to approach the general problem of representing continuous systems and processes. Based on the ground-breaking work of Forester (1966) and the contributions of Richardson and Pugh (1983) and Hannon and Ruth (1994), systems dynamics is a methodology for understanding problems that are (1) dynamic in that they involve quantities that change over time, and (2) characterized by active feedback effects. Thus, this modeling approach to problem-solving applies to dynamic (continuous) problems that develop within systems characterized by feedback: a quality that has been shown to apply to both human systems and a broad spectrum of environmental processes (Hannon & Ruth, 1994; Ford, 1999).

Successful application of dynamic modeling hinges on an understanding of feedback concepts and their role in defining dynamic systems. Perhaps the most basic definition of feedback explains the concept as the transmission and return of information to a system. This simple definition is central to developing the causal thinking needed in order to organize ideas in a system dynamics study (Roberts et al., 1983). Using feedback as a structuring concept, causal thinking requires the model builder to isolate key factors that direct the processes involved and explain their relationship to the system of interest. Through this type of mental experimentation the logic and connections that explain some observed behavior could be diagnosed. Illustrating these elements and connections in a simple systems diagram helps to reveal the causal influences that form the system and the possible feedback effects that may be at work (Fig. 7.5).

To demonstrate the general approach to dynamic modeling we can explore the farmland conversion process as an example. In this simple model we are interested in understanding the mechanisms that contribute to the change in land use from farmland to urban. At the rural–urban fringe of our hypothetical planning area, land market forces and the personal motivations of private landholders coupled with local government growth strategies have contributed to the gradual but steady conversion of land from agricultural uses to urban. While this example cannot approach the level of sophistication needed to

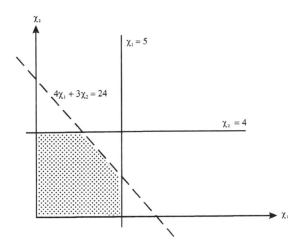

Fig. 7.4 Defining decision regions using the simplex method.

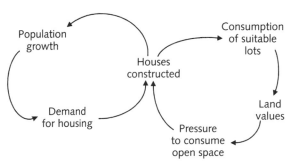

Fig. 7.5 Causal loop diagram of the urban development process.

explain the process completely, we can create a model that allows us to explore a set of plausible relationships that can be used to draft a more comprehensive farmland preservation program. Given the observed pattern indicative of the decline in farms in the region, we can create a scenario that can be evaluated by the model.

1 Unmarried young adults migrate to regional centers for better job opportunities.
2 Population decreases and births fall below replacement.
3 Reduction in business services due to declining demand.
4 Reduction in key social services.
5 Loss of services induces out-migration of young families and ensures in-migration is minimal.
6 Aging population contributes to disintegration of balance in the community.
7 Farmholdings passing to next generation are offered for sale instead.
8 Farmland parcels are purchased by land speculators and held out of production.
9 Land costs within the urban center encourage decentralization of population and commercial activities.
10 Land held in speculation is sold for development.
11 Development increases market value of existing agricultural lands.
12 Increasing prices and taxes encourage marginal producers to sell land.

In this example, 12 causal forces were identified that contributed to the general process of rural to urban land conversion and the decline of farmland. Each of the relationships expressed can be reviewed to determine how well they explain the observed behavior. Once a reasonable sequence of cause and effect is achieved, the relationship can be specified and diagrammed. Diagramming causal relationships enables the chains and loops that connect elements of the process to be visualized. In addition, diagramming facilitates development of the model's initial structure and assists with the representation of feedback effects. The causal-loop diagram representing the farmland conversion process is shown in Fig. 7.6. Following the looping arrows illustrated in the figure gives the impression of causality, suggests how this system may behave over time, and defines the nature of feedback in the system. While this form of open-loop thinking is a central feature of dynamic modeling, understanding the behavior of feedback is the ultimate goal of the systems dynamic approach.

Feedback can assume several forms in a dynamic system, including:
• Single-loop positive feedback
• Single-loop negative feedback
• Multiple-loop feedback.

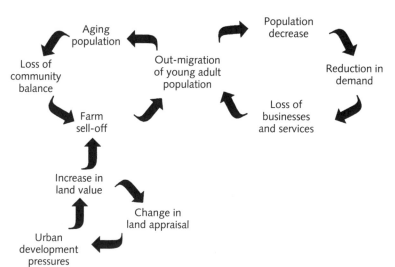

Fig. 7.6 The farmland conversion model.

With a focus on these feedback processes, dynamic modeling proceeds on the assumption that feedback structures are responsible for the changes a system undergoes over time, the premise here being that dynamic behavior is a consequence of system structure. Therefore, as both a cause and consequence of feedback, dynamic modeling looks within the system for the sources responsible for its behavior.

As suggested by Richardson and Pugh (1983), any problem viewed from this perspective is likely to be seen initially as a graph of one or more variables changing over time. Such graphs assist the modeling processes by

1 Focusing attention on the problem.
2 Helping to identify key variables in the system.
3 Helping to define the problem dynamically.

Graphing important variables and inferring graphs of other significantly related variables produces the problem focus of undertaking this type of study. In essence, displaying graphs over time is the reference mode of behavior for the model and gives limited indication of the patterns that will develop and that should be incorporated into the modeling effort.

The dynamic model takes form as the variables defined by the causal loop diagrams are translated into the structural components that will become the formal model. Translation requires representing each component of the system according to the role it plays in the process. According to the language of dynamic modeling there are four critical components used to construct a model:

- Stocks or levels – defining the value or accumulated value of a variable.
- Flows or rates – explaining process equations that act on the stock or level.
- Connectors – describing links that transfer information about the stock or level to control.
- Time step – defining the period of time over which the values of stock should be updated during a run of the model.

Typically, translation involves transferring the model from its representation as a causal diagram to a computer model using a special-purpose simulation language. Presently the commercially available simulation program developed by High Performance Systems called STELLA II typifies the characteristic dynamic modeling environment. Programs such as STELLA II provide a range of functionality for quickly prototyping computer simulation models. The main advantage of this type of modeling environment is that it does not require the designer of the model to possess knowledge of a conventional programming language such as C, Pascal, or Fortran. Modeling tools like STELLA II are a type of graphical programming language that employ graphic objects which can be assembled on a "white board" to construct a functioning system model. Placed into this design construct the form of the model can be easily understood and the processes acting on the problems can be more quickly identified. However, the greatest advantage of this approach is the relative ease by which changes can be made to the values given to variables in the model. Since the values assigned to variables and constants can be modified, exploring contrasting scenarios, adjusting time horizons, and altering functional relationships can be done without altering the general structure of the model. This feature greatly enhances the role of dynamic modeling as a decision support tool.

Cellular automata

Cellular Automata were introduced in the late 1940s by John von Neumann (von Neumann, 1966; Toffoli, 1987) and popularized in the late 1960s with the development of the "Game of Life" (Gardner, 1970; Dewdney, 1990). Cellular automata are often described as the counterpart to partial differential equations, which have the capacity to describe continuous dynamic systems. A cellular automaton is essentially a model that can be used to show how the elements of a system interact with each other. The basic element of a cellular automaton is the cell. A cell is a type of memory element and stores states that represent characteristics of the system under investigation. These cells can be two-dimensional squares, three-dimensional blocks, or they may take some other geometric form such as a hexagon. Each element comprising the system is assigned a cell and cells are arranged in a configuration. For example, cells joined together to form a single line comprise a one-dimensional cellular automaton, whereas cells arranged on a grid form a two-dimensional cellular automaton. In either instance, cells arranged according to a one-dimensional or two-dimensional lattice represent a static state. To introduce change (or dynamics) into the system, rules must be added to the model. The purpose of these rules is to define the state of the cells for the next time step. In cellular automata, dynamics occur within neighborhoods, and different definitions of neighborhoods are possible. For the two-dimensional lattice four neighborhood definitions are common: (1) the von Neumann neighborhood, (2) the Moore neighborhood, (3) the extended Moore neighborhood, and (4) the Margolus neighborhood (Fig. 7.7).

In the initial configuration of the cellular automata, each cell is assigned a "starting" value from the range of possible values typical of the system under study. For instance, if the range of possible values (states) were 0 to 1, then each cell would be assigned a 0 or 1 in the initial configuration. A transition function and a transition rule are also associated with each cell. Working in concert, the transition rule and function take as input the present states (values) of all the cells in a given cell's neighborhood and generate the next state of the given cell. When applied to all of the cells individually in a cellular automaton, the next state of the whole cellular automaton is generated from the present state. Then the next state of the cellular automaton is copied to the present state and the

process is repeated for as many clock cycles as desired.

Cellular automata are rapidly gaining favor as a tool for modeling dynamic spatial systems (Batty & Xie, 1994; Cecchini & Viola, 1992; Engelen et al, 1996; White & Engelen, 1997). When compared to traditional approaches based on differential or difference equations, cellular automata have notable advantages: they are inherently spatial with rule-based dynamics; computationally efficient; can model systems with very high spatial resolutions; and provide and intuitive link to geographic information systems data formats (White & Engelen, 1997). For these reasons cellular automata show great promise as a basis for regional and environmental modeling (Clarke et al., 1994; White & Engelen, 1997). For example, in a recent study undertaken by Clarke, Hoppen, and Gaydos (1997), a cellular automaton simulation model was developed to predict urban growth as part of a project for estimating the regional impact of urbanization on climate. In this study the rules of the model were more complex than for the typical application, and allowed specific growth scenarios to be performed by the model using historic land-use/land-cover data sets. A similar study conducted by White and Engelen (1997), an integrated model of regional spatial dynamics, consisted of a cellular automaton-based model of land use linked to a geographic information system, to nonspatial regional economic and demographic models, and to a simple model of environmental change. Based on their initial testing of this integrated package, the cellular automata

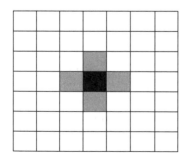

Von Neumann neighborhood

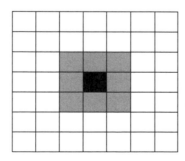

Moore neighborhood

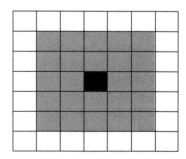

Extended Moore neighborhood

Fig. 7.7 Cellular neighborhoods.

enabled detailed modeling and realistic prediction of land-use patterns and provided a way to introduce environmental factors into the simulation. Although the authors caution that it may not be possible to predict the state of any land-use system far into the future, reasonable forecasts of land-use patterns over a period of 10 to 15 years can be made with measured confidence if the growth rate of the region is known. What this model does is support "what if" experiments, allowing the user to explore various possible futures and develop insights that may be of use in strategic planning.

Summary

As a future-oriented activity, environmental planning relies on the application and development of models to explore "what if" situations and test the outcome of various policy decisions and alternatives. The issues related to the use of models in environmental planning and the design of simulation experiments were examined in this chapter. Introduced in this discussion were the foundation techniques that rest at the core of many of the models used in environmental planning. The techniques selected for this discussion included Monte Carlo sampling, Markov chain analysis, dynamic modeling, and cellular automata. With a solid understanding of these techniques the advantages and disadvantages of using models in planning can be understood. In addition, the methods of modeling outlined in this chapter help the environmental planner decide what to model, how to design a simulation experiment to acquire information, and how to interpret the results of modeling studies. Taken together these topics support the intelligent use of models and stress their value in the formulation of environmental plans.

Focusing questions

What is a model and how do models contribute to the simulation process?

Develop a simple scenario: outline the causal chain of events that might unfold following a decision to relocate and expand a highway through a rural area.

Compare and contrast dynamic, probabilistic, and stochastic models; how do these forms influence the representation of a system?

What considerations guide the use of models in general, and how do they influence the interpretation of results obtained from a simulation experiment?

CHAPTER 8

The Decision Support Perspective

Effective environmental planning demands continuous and reliable sources of information to assist in identifying problems, establishing priorities, and formulating plans. Without question, a primary activity of the planner involves "sensing" the planning area, taking in data on critical indicators of the regional system, analyzing that data, and applying the information derived from it in order to react appropriately to the needs of the community and its environment. A technology has emerged that supports the acquisition of data and guides its transformation from raw numbers and facts to meaningful information systems. This information technology is based on the computer and its ability to aid environmental decision-making (Lein, 1997). In this chapter the computer and its application in environmental planning will be examined. Beginning with a review of essential information technology concepts and information management, this chapter will explore the nature of computer assisted planning and role of geographic information systems in decision support.

The role of information technology

The planner's need for timely and reliable information concerning the planning area is not a recent phenomenon. As suggested by Kaiser et al. (1995), information represents a "strategic

intelligence" that not only enables the planner and community to identify, understand, and deal with new and vexing situations, but to do so systematically in a manner that purposefully compiles, organizes, and analyzes data to realize how the community and its environment has changed. Throughout human history and certainly since the industrial revolution the ability to acquire this "strategic intelligence" has moved society from clay tablets to digital databases capable of moving information at near light speed almost anywhere over the globe. Although the volume of information available to us has grown substantially, the innate human capacity to process this information has remained essentially the same. Therefore, while having information is important, more critical is the ability to use this information and generate answers from it. Taking full advantage of the strategic potential of information has fostered the creation of new approaches to accelerate accessibility to information, enhance its processing and analysis, facilitate its storage and utilization, and perform these tasks quickly and reliably.

Information technology identifies the range of approaches that have been developed to assist with the gathering, storage, production, and dissemination of information in order to meet the needs of complex decision-making. To illustrate this point we can turn to the example provided by Lein (1997). Fifty years ago a planner seeking guidance regarding either the resource potential of land, its developmental constraints, or its general characteristics was unable to proceed without extensive field surveying and mapping followed by weeks of manual data preparation and analysis. At the dawn of the twenty-first century a satellite can gather quantities of data in a single pass over the planning area that can easily overload the cognitive abilities of its users and saturate the decision-making process. Consequently, there is a need to refine and simplify data and to produce information that can feed the decision-making process. However, information technology is not only important in its ability to store and refine information, but also by its ability to support those who use it by allowing information to be used in new ways. The challenge is that while information technology may be an integral part of planning, planners must to able to connect technology to the task and responsibilities of day-to-day practice (Moffat, 1990).

Connecting information technology to the planning process begins with a basic appreciation of the information carried by this technology and the processes that generate it. Each of us uses information continuously in our daily lives, assessing conditions and making decisions, yet most of us would be hard-pressed to define precisely what is meant by the terms or to describe how information is used (McCloy, 1995). Debons et al. (1988) have offered a simple way to categorize information in a way that relates to environmental planning:

- Information as a commodity – recognizes that information can possess economic value that will influence who controls and disseminates it.
- Information as communication – describing the condition where the exchange of information transfers understanding and meaning.
- Information as fact – explaining the general state where data devoid of context conveys no relevant meaning.
- Information as data – identifies the product of symbols organized according to established rules and conventions.
- Information as knowledge – underscores the intellectual capability to take information and with it extrapolate beyond simple facts and data to draw meaningful conclusions.

From these contrasting definitions, we can see that the term "information" can be applied to a wide range of cognitive states that suggest a range of differing functional roles. A technology dedicated to the management of information must therefore preserve critical aspects of these conditions, particularly if information is to remain useful and maintain its relevance as a resource to those who rely on it (Lein, 1997). In a slightly different context, we can also describe information as a progressive entity that begins with data and is sequentially transformed into more meaningful states of "being." Each of these states carries the implicit assumption that transformation enhances knowledge and understanding of the problem. Expressed in this way, information is one stage in the continuum that starts with data and ends with knowledge (Fig. 8.1). An information technology that "manages" information must also mirror aspects of this continuum, connecting the purely data-driven processes of information access and flow to the highly cognitive activities that envelop knowledge and wisdom (Lein, 1997).

From data to information

For the purposes of environmental planning, we may define data as the raw (unrefined or processed) observations about objects, events, or surfaces that comprise the planning area. This notion introduces the "layer cake" model as a means of conceptualizing the planning area as a series of data planes each conveying a specific theme of relevance to the planning. While such data may, at times, be used directly as information, to be useful this data must be refined or pro-

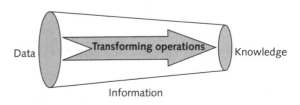

Fig. 8.1 Translating data into knowledge.

cessed in some way to make it more useful for problem-solving.

The rationale for processing data in order to create information has been summarized by McCloy and includes two important ideas. First, data pertaining to the physical attributes of the environment may be related to, but usually does not constitute the actual parameters, required by the planner. Secondly, data may be far too detailed or complex to use in its present form. Therefore, by transforming data into information, the "knowledge base" needed to guide planning decisions is created. Information technology functions to facilitate data transformation and analysis and specifically addresses procedures related to:

- Generalizing from data.
- Estimating physical parameters through data transforms.
- Filtering data to identify trends and anomalies.
- Modeling relationships in data over time and space.

Through the application of these procedures information is derived that can serve as the building blocks of a solution. Therefore, to have value, information must either enable decisions or facilitate subsequent actions by the environmental planner.

Information technology is a term loosely applied to any automated system that enhances the value of information. Generally, these technologies orbit around the nucleus of a data–information–knowledge approach to environmental planning, and define techniques instrumental to the task of representing data, information, or knowledge with an inherently geographic component. Although the techniques that form an information technology will change according to the pace of change of technology in general, their role remains constant: to support the needs of the decision-making process.

Planning and decision support

The concept of computer-aided decision support was discussed nearly two decades ago by Langendorf (1985). The basic idea then, as now, involves the design of computer systems that can support

and automate critical aspects of the decision-making process. Although a simple definition of this concept has yet to be produced, a decision support system targets situations where the planner is forced to confront problems that are poorly structured or ill-defined. Problems of this variety are typically unique in the sense that they display characteristics that are highly variable, complex, and contain a high level of uncertainty. Examples might include problems encountered when conducting environmental impact studies, risk assessments, or other activities where there are a lot of different factors to consider and nothing is absolutely clear. As a consequence, unstructured problems cannot be addressed using standard operating procedures. Rather, the planner when presented with an unstructured problem relies heavily on intuition, judgment, prior knowledge, and adaptive problem-solving behavior to compensate for the extremes of uncertainty that punctuate the situation. Like the examples of environmental risk and hazard assessment, facility siting, or environmental impact assessment, there is a need to help focus judgment, support intuition, apply prior knowledge, and create an environment where adaptive strategies can be directed toward defining, analyzing, and evaluating problems. Developing an automated system to support decision-making involves the identification of tools that can supply timely and accurate information to improve the decision-making process. Therefore, the motivating force behind the concept of decision support and the development of technologies designed to aid the planning process is essentially the goal of helping planners make better decisions.

"Decisioning" and geographic information systems

Developing technologies to support decision-making begins by taking the decision-making process apart to reveal its fundamental structure. For the purposes of planning support, we can simplify the decision-making process by reducing it to its three root components:

1 Acquiring, retrieving, and selecting relevant information.

2 Structuring the decision problem to enhance the visibility of alternatives.

3 Evaluating alternatives based on their relative attractiveness.

Accepting the premise that decision-making can be reduced to these basic activities, decision support becomes a methodology that can be developed and applied to each task (Vlek et al., 1993). In this context decision support becomes a means to channel and direct the planner's cognitive processes in order to reduce uncertainty. Given the realization that most of us are limited and selective information processors capable of making simple, adaptive decisions, but poor at making complex, strategic decisions, the role of decision support becomes clear: to overcome cognitive limitations and extend the decision-maker's informal reasoning and evaluative capacities.

Understanding the types of problems where decision support can be applied is critical not only to the design of decision support systems, but also to their effective use. Five categories of problems arranged according to their relative level of difficulty have been presented by Davis (1988) (Table 8.1). In general, decision support is most effective when the problem is complex or ill-defined. Examples of such problems include situations where:

1 Objectives are difficult to determine or are conflicting.

2 Alternative actions that might be taken are difficult to identify.

3 The effect of an alternative on a given outcome (result) is uncertain.

In these instances decision support is targeted at the "structurable" part of the problem. This focus paves the way for the design of "tools" to assist problem structuring.

Automated decision support defines a computerized system that incorporates functionality to collect information, formulate problems, and perform analysis to help decision-makers address situations that are ill-structured. As shown previously by Lein (1997), information collection in a decision support system involves the process of gathering data and information from the user of the system and from a repository of pertinent data. Typically this takes the form of a database. Prob-

Table 8.1 Levels of decision problem.

Problem type	Description
Type O	Mechanistic, non-thought-provoking problem with few alternatives.
Type A	Slightly more complex with more alternatives to consider; automation helps illuminate problem.
Type B	Requires "brute force" techniques to find best course of action, since number of objectives and possible choices are large.
Type C	Problems where the structural complexity and sheer size of the problem reduce ease of problem visualization.
Type D	Dynamic and extremely complex problems that support only a qualitative "hit or miss" approach.

lem recognition describes the process of refining the initial definition of the problem through a variety of data visualization techniques, data exploration tools, and models that can be embedded in the system. Analysis undertaken using an automated decision support system can mean many different things. For the most part, analysis suggests the application of specialized tools or routines to connect the collected data with "models" that aid prediction or explanation. Based on these characteristics, it can be seen that decision support systems are specifically oriented toward the types of information-processing needs demanded by the decision-making process. Thus, armed with an automated support environment, the environmental planner can manipulate large volumes of data, perform complicated computations, and investigate relationships in data that might otherwise go unnoticed. This has become the role of a geographic information system (Fig. 8.2).

A geographic information system (GIS) can be defined in several ways (Maguire, 1991). Three common views of this critical information technology include:

1 Toolbox-based definitions that characterize GIS as a powerful set of tools for collecting, storing, and retrieving geographically referenced data, and transforming and displaying that data to reveal new patterns

Attribute data

Spatial data

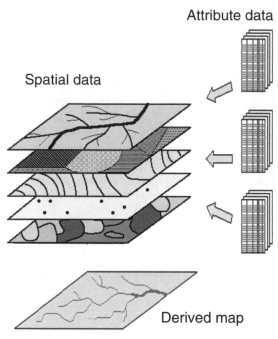

Derived map

Fig. 8.2 Spatial data representation via GIS.

based on real world relationships (Burrough, 1986).

2 Database definitions that explain GIS as a database system in which most of the data is spatially indexed and upon which a set of procedures operate to answer queries about geographic entities in the database (Smith et al., 1987).

3 Organization-based definitions that describe GIS as a decision support system involving the integration of spatially referenced data in a problem-solving environment (Cowen, 1988).

Still others consider GIS a paradigm consisting of a collection of information technologies and procedures for gathering, storing, manipulating, analyzing, and presenting geographic data and information (Huxhold & Levinsohn, 1994). Although these descriptions of GIS are useful because they help us place the technology into a "problem" context, when a GIS exists in an operational setting, such as a planning department or governmental agency, we can consider GIS from an entirely different perspective.

Removing GIS from direct consideration of the software systems and hardware environments that most commentators dwell upon when discussing this technology, and focusing instead on how GIS functions within the context of decision-making, allows an alternative explanation of GIS to emerge. This considered view recasts GIS less as a technology and more as a methodology. The methodological perspective carries important implications to environmental planning and other disciplines that incorporate GIS in their day-to-day operations. As a methodology GIS represents a form of "geographic" thinking that influences how problems are conceptualized, how data is organized, how information is generated, and how knowledge is applied. Because the central feature of this methodology is its geographic focus, the manner by which spatial information supports discourse and decisions regarding population, economic, land-use, environmental, and other patterns that constitute the planning area takes precedence in formulating and analyzing planning problems. For example, if we consider the suitability assessment problem introduced in Chapter 5, applying GIS would allow the planner to identify the objectives of the problem carefully by permitting a visual examination of the factors involved, assembling the data in a "model" that could be used to create an expression of suitability, and perform an analysis that could be displayed effectively in the form of a map (Figs. 8.3a and 8.3b).

Therefore, using GIS to guide planning builds critical knowledge and support into the process by

• Describing the history and current status of critical planning variables.
• Forecasting their future status.
• Monitoring, mapping, and interpreting how these variables change.
• Diagnosing and planning, and development problems.
• Modeling critical relationships and impacts.
• Presenting information to policy-makers and the public.

GIS methodology capitalizes on representing problems and revealing solutions in a highly visual way. Examples include maps, charts,

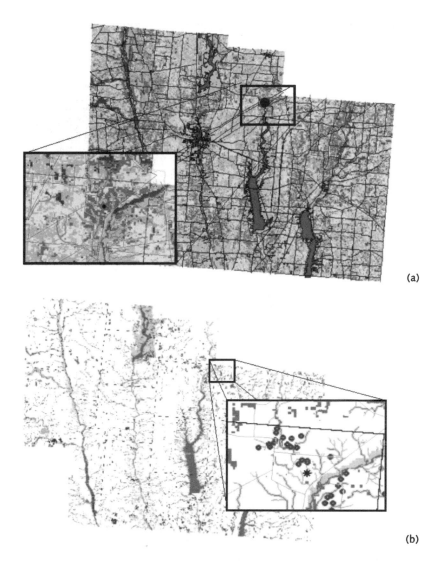

Fig. 8.3 (a) GIS-derived developmental suitability map. (b) GIS-derived critical resource map.

diagrams, and other illustrations that permit the geographic nature of most planning problems to be visualized and examined (Fig. 8.4).

Placed into a problem-solving or policy-making role, GIS displays simplify and abstract information, conveying that information more effectively than numerical tables or written narratives. Display also offers the chance to create alternative views of information that give new insight into the spatial and contextual features of the planning area. A geographic information system has a unique but under-utilized ability to examine data and reveal associations or correlations among factors that are inherently geographic (O'Looney, 1997). This ability to "see" spatial associations gives the planner that opportunity to define problems in new ways and analyze multiple spatial data-sets. Users of GIS can pose questions via the system and operate on data. Such questions can then be taken to the next level and placed within a future context. These "what if" questions become a useful way to explore the spatial footprint created by various planning alternatives. Here, GIS can be envisioned as a conduit that not only provides a means to organize and access spatially referenced data, but also as a platform that

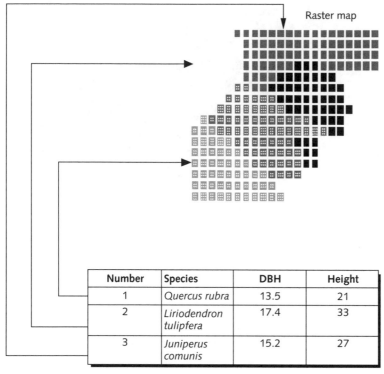

Raster map

Number	Species	DBH	Height
1	*Quercus rubra*	13.5	21
2	*Liriodendron tulipfera*	17.4	33
3	*Juniperus comunis*	15.2	27

Fields

Fig. 8.4 Fundamental GIS data model.

encourages cross-disciplinary thinking and information sharing. In professional disciplines such as environmental planning, the "common ground" GIS provides encourages broader solutions to complex problems that allow specialists in transportation, civil engineering, geology, sociology, and others to interact. Yet, as with any applied methodology, credible answers derived from a GIS depend upon

- The integrity of the user.
- The skill of the user.
- The integrity of the data and methods used.
- Compatibility among the data sources that are being integrated into the analysis.

Developing GIS capabilities that facilitate environmental planning efforts draws on the functional capabilities of this technology. Common functions that GIS performs are illustrated in Figs. 8.5a and 8.5b and include map overlay, buffering, and specialized operations such as viewshed analysis, surface analysis, and short path analysis. By carefully considering these func-

Table 8.2 General benefits of GIS.

Improved quality of information
Improved timeliness of information
Enhanced information flow
Increased productivity
Reduced costs
Improved decision-making

tional capabilities, the net benefits of GIS can be realized (Table 8.2). When these general benefits are examined in relation to the goals of environmental planning, the greatest potential for GIS lies in its ability to integrate key public values in weighing solutions to community problems (O'Looney, 1997).

From a local government perspective this suggests that GIS applications are driven by a deep concern for:

- Efficiency – explaining the relation between the amount of effort and the amount of return.

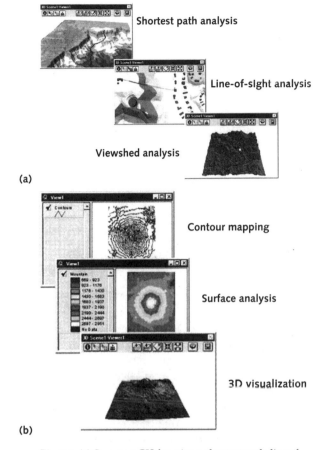

(a)

(b)

Fig. 8.5 (a) Common GIS functions: shortest path, line of site, and viewshed analysis. (b) Common GIS functions: contour mapping, surface Analysis, and visualization.

- Equity – describing the relationship between a citizen's status or effort and the social benefits they receive.
- Community viability – defining characteristics such as civic participation, cultural institutions, and activities central to a working social order.
- Environmental quality – identifying qualities of the ecosystem that provide for the long-term maintenance of life.

The manner in which GIS is used to balance these concerns has been reviewed in detail by O'Looney (1997). Overall, the success of GIS is highly dependent upon the factors that contribute to its satisfactory implementation. With respect to environmental planning, successful implementation is guided by the attention given to:

- System design and purpose.
- Data acquisition and database development.
- Analytic functionality and error management.
- Long-term management, maintenance, and modification.

Although our principal interest in this discussion is the realization of GIS as a problem-solving methodology, to understand how this method can be applied and to appreciate the significance of GIS design and implementation, a review of the fundamental components of a system must be entertained.

GIS design for environmental decision support

Geographic information systems have three important components: (1) the physical GIS comprised of computer hardware; (2) the functional GIS – consisting of sets of application software models; and (3) the organizational GIS – describing the decision-making environment in which GIS is implemented (Fig. 8.6). Each of these parts is equally important and needs to be balanced if the total system is to function according to the philosophy that encouraged its creation and design (Burrough & McDonnell, 1998).

The physical GIS

The hardware environment of the GIS describes the computer system and the related peripheral devices required to input, process, and output geographic information. While the hardware component of the GIS is extremely dynamic, with advances in computer technology introducing new capabilities in rapid succession, several devices are common:

1 **Data input devices** – such as digitizers or scanning systems that enable geographic information to be captured from its analog form as a map and recorded as digital data in the computer.

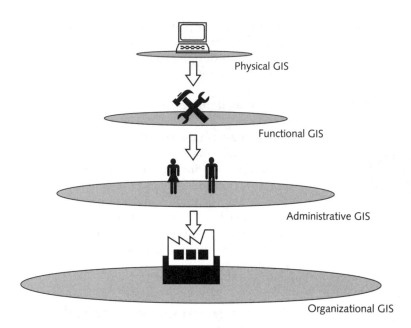

Physical GIS

Functional GIS

Administrative GIS

Organizational GIS

Fig. 8.6 The levels of GIS.

2 **Output devices** – such as printers, plotters, or film recorders that provide hardcopy products of the results generated by GIS analysis.

3 **Storage devices** – such as fixed disks, floppy disks, CD-roms, or tape systems that permit digital geographic data to be written and organized into files.

The functional GIS

The software component of GIS characterizes its functionality. The functional qualities of GIS can be divided into five main categories:

1 **Data input and verification** – describing a collection of software routines that guide and manage the capture of geographic information and their conversion to a standard digital form.

2 **Data storage and management** – involves the database management system that implements a specific spatial data storage paradigm (raster, vector, quadtree) and the functionality needed to control, organize, and update this database.

3 **Data output and presentation** – explains the various routines that direct the creation and display of maps, reports, and other results generated from an analysis. In a GIS information may be presented in a variety of ways including transfer to print of plotting devices and well as the direct conversion to numerous graphic file formats (gif, jpg, tif).

4 **Data manipulation and transformation** – functionality in a GIS related to this class of operations explains a set of routines that guide the removal of error and subsequent updating of data together with routines designed to perform geographic analysis. The analytical toolbox of the GIS can be rich with routines that perform data modeling, measurement, logical retrieval, and logical combination. Sets of common analytical requirements that typify GIS functionality are given in Table 8.3.

5 **User interface** – recognizing the application-specific nature of GIS, most systems provide some ability to customize and enhance how the user of the GIS interacts with the software environment. Presently most systems operate using a command language interpreter or by means of a graphical user interface (GUI). With either mode of interface, systems can be expanded functionally through the use of

Table 8.3 Typical GIS functionality.

Maintenance and analysis, spatial data	Format transformations Geometric transformations Map projections/transformations Conflation Edge matching Editing graphic elements Line thinning
Maintenance and analysis, attribute data	Attribute entry and editing Attribute query functions
Integrated analysis functions	Data retrieval, classification Measurement Overlay operations Neighborhood operations: • search • interpolation • regionalization Connectivity functions: • proximity • intervisibility • network • spread
Output formatting	Map annotation Text labels Symbolization

Table 8.4 Stages of GIS implementation.

Stage	**Tasks**
Needs assessment	Conduct interviews and examine current operational tasks, projects, and information products
Requirements analysis	Specify system philosophy, conceptual design; identify technical constraints
System design	Specify design requirements, determine optimal data model, craft prototype system
Outsource	Generate requests for proposals and evaluate, select vedor(s), establish benchmark tests, evaluate system performance
System construction	Develop pilot project, re-evaluate design, develop data encoding and conversion schedule, implement quality control program
Evaluation and enrichment	Re-evaluate needs, add capacity, modify design

simplified, formal programming languages. These macro-languages can be used to tailor GUI displays or link basic application commands together into a single command or command sequence.

The organizational GIS

When the five elements described above are assembled together they form a GIS platform that is then placed into an administrative setting. The administrative setting oversees the management, maintenance, and support needed to keep the GIS functioning according to its designed intent. Establishing this level of administration and specifying the goals of the system embodies the organizational setting in which GIS operates. This organizational context defines the arena where decision-making occurs and the structure in which GIS lends its support. At this level, concern is directed away from the physical and functional aspects of the system and attention is given to the overall rationale for using GIS, and the

specific purpose and philosophy that define the system's "reason for being." It is typically at this level where GIS design and implementation begins.

A successful GIS is created to fulfill one or more goals of the organization that recognizes the advantages of utilizing "geographic" data to support decision-making. Rather than seeking a "magic bullet" in a single measure of evaluation, a practical solution involves a design process that incorporates all the concerns that influence GIS implementation (Chrisman, 1997). The general sequence of steps that direct this process is outlined in Table 8.4. Within this process the concept of "geographic" data stands as the central feature of design and the creation of a functional GIS. Geographic reference gives data new meaning by relating a name, value, quality, or condition to an object that can be visualized and treated as an entity that occupies real geographic space.

GIS design can be approached in either of two ways. One design path is referred to as the focused approach. Focused GIS design concentrates development around a well-defined and clearly articulated application or purpose. Typical applications that may encourage focused design include developing a GIS for hazard assessment and emergency management, transportation planning, growth management, or zoning administration. In each of these examples a relatively high-priority problem has been identified. With design directed toward a relatively high-priority problem or need, decisions concerning database design, functionality, and management are all predetermined by what is required to address the problem. An alternative design strategy is referred to as the panoramic approach. The panoramic approach to design is not focused on a single problem but instead envisions GIS in a broad context where the methods of GIS analysis can be applied to a range of applications. A GIS developed according to this strategy functions primarily as a dynamic store of data where the user community interacts with a limited functionality that can be expanded as the system matures. Panoramic design, therefore, creates an open-ended system that maintains a purposely broad view of GIS that allows the system to evolve in a modular fashion.

While the approach to design may vary, each requires a sustained commitment throughout the GIS lifecycle. That commitment develops out of a long-range problem whose solution, or some part thereof, can be obtained from the information derived from the GIS. From this point forward the specifics of design fall into place. Here designing the GIS involves addressing questions pertaining to:

1 The informational needs the system will be called upon to deliver.
2 The data needs users of the system will require in order to perform their jobs.
3 The software functionality that will be employed to deliver the required informational products.
4 The hardware capabilities needed to ensure optimal software and database performance.

With these considerations understood, design efforts can be directed toward the issues of data acquisition and the overall architecture of the database. The fundamental stages followed in developing the GIS database are identified in Table 8.5. As suggested by Table 8.5, development involves three primary tasks:

- The assessment of informational needs.
- The collection and evaluation of relevant data sources.
- The specification of a conceptual database design.

Because the "geographies" that comprise the planning area are inherently complex, they must be abstracted and structured into a spatial data model that not only serves as a formal device for representing their essential characteristics, but also provides for an efficient means of storage within the computer. This spatial data model is then translated into a data structure that is then encoded into the appropriate file format. Presently, GIS technology recognizes two principal ways of representing spatial data: (1) raster format and (2) vector format. Although it is common to convert data between these two structuring paradigms, during the design process the question of which mode to employ often raises confusion. Useful guidance has been offered to help resolve this issue (Burrough & McDonnell, 1998) (Table 8.6).

With the data acquisition and database design issues resolved, consideration can be given to the functional capabilities of the GIS and the analytic operations it will be called upon to perform. Approaching the question of functionality requires

Table 8.5 Stages in GIS database design.

Phase 1 – Information needs assessment
- conduct informational interviews
- review existing documents
- determine the planning area
- analyze information needs

Phase 2 – Data evaluation and collection
- collect pertinent data
- assess existing data coverage
- evaluate data quality

Phase 3 – Database specification
- develop data standards and classification schemas
- select optimal scale and resolution
- establish data input and update schedules
- design file/map manuscript system

Table 8.6 Comparison of raster and vector data models.

Raster	Vector
Advantages:	
Simple data structure	Good representation of entity data model
Location-specific manipulation of data	Compact data structure
Supports wide range of analysis	Explicitly described topology
Ease of modeling	Ease of coordinate transformations
Inexpensive technology	Accurate graphic representation
Disadvantages:	
Large data volumes	Complex data structure
Loss of information within cells	Analytic complexity
Crude graphic display quality	Simulation and modeling operations
Coordinate transformations difficult	more difficult

the adoption of a "toolbox" view of GIS. As a toolbox, GIS contains an array of routines or algorithms to process and manipulate spatial data. The tools needed to apply GIS in environmental planning have been critically reviewed by Kliskey (1995). In general, GIS software consists of two integrated components:

1 A core module of basic mapping and data management routines.
2 An application module that consists of a menu of geographic analysis routines.

Because the analysis of geographic patterns and associations is the driving force behind the use of GIS in environmental planning, the ability of the system to perform geographic analysis and model fundamental geographic concepts is critical. While nearly all software systems provide tools for collecting, organizing, and displaying spatial data, their ability to perform complex analytical operations on spatial data sets varies. In the majority of cases the toolbox of a GIS consists of routines that perform overlay, dilation (buffering), neighborhood operations, reclassification, query, distance operations, tabular analysis, and display. A detailed listing of common GIS functions is provided in Table 8.7.

Functionality together with the database plays an important role in specifying the design of the GIS. Of equal importance, particularly with respect to the long-term success of the system, are those factors that influence management and maintenance. Since effective planning demands accurate and timely information, capitalizing on

Table 8.7 Common GIS modeling functions.

Function	Description
Area	Measures areas associated with data
Cluster	Performs cluster analysis on data set
Cover	Superimposes one layer on top of another
Distance	Measures Euclidean distance
Euler	Determines shape or form of features
Extract	Extracts values from one layer to another
Filter	Performs data smoothing
Group	Determines contiguous groups
Histogram	Computes frequency histogram
Interpol	Interpolates continuous surface from point data
Overlay	Performs Boolean combinations
Pathway	Finds shortest path through network
Reclass	Classifies layer attributes into new categories
Surface	Performs slope and aspect calculations
Trend	Conducts polynomial trend surface analysis
View	Performs intervisibility analysis

GIS as a centralized source of data required a level of management to ensure that the system performs in a manner that keeps pace with the demands placed on it. Anticipating and planning for change is central to GIS, particularly as the system becomes operational. Four maintenance tasks are ongoing features of the GIS and should be considered early in the design process:

1 Database updating.
2 Software revisions and enhancements.
3 Hardware upgrades.
4 Application refinements and expansions.

As GIS is a data-driven environment, the accuracy, fidelity, and completeness of the database will determine how well GIS lends support to planning. The planning arena, however, is dynamic, which means that any database runs the risk of losing its utility. Thus, a detailed program must be created to plan and schedule the updating and revision of the GIS database. The specifics of this plan depend heavily on factors such as:

- The rate at which new data becomes available.
- The planner's need for timely information.
- The pace of change characterizing the planning area.

The pace of change is also a determining factor with respect to GIS software. As new innovations in information technology appear, new developments in GIS software can be anticipated. A strategy is therefore needed to help focus questions pertaining to how and when to adopt software revisions, and to critically evaluate software enhancements through detailed benchmarking to avoid adding functionality that may be unnecessary, have limited utility, or fail to support the goals and purpose of the system. A similar strategy is needed to deal with the bigger problem of hardware upgrading and replacement. In general, keeping pace with technological progress involves carefully identifying the "appropriate" level of technology needed to drive the system, maintain acceptable performance standards, meet the continuing goals and objectives of the operational GIS, and provide timely support for planning. Finally, the question of application refinement can be approached. Given the likelihood that GIS when first implemented was designed for a limited number of application areas, the need will develop to broaden the methodologies imported to the system. Here, as planners discover how GIS supports analysis, existing applications may be refined, streamlined, and customized through the creation of macros and other devices that enhance productivity. Refinement can also contribute to experimentation with new problem-solving approaches tested for GIS feasibility that may lead to their eventual adoption in day-to-day practice. Integrating GIS into the planning process and capitalizing on the support capabilities it offers is examined in the following section.

GIS-guided planning support

When connected to the planning problem, GIS defines a methodology whereby the tasks and analytical needs of the planner find their solution in the combination of data and the functional capabilities of the system. Once GIS has been installed and its components implemented, the main focus of concern shifts from design and development to the issues surrounding the system's operational use. In an often-cited study undertaken by Campbell (1994), it was shown that in the UK there was a surprising under-utilization of GIS in planning agencies. Although GIS was widely applied to automate map-making, there was comparatively little use of complex spatial analysis functions. While several reasons were offered to account for this disparity in application rigor, a major factor involved the general problem of learning how to "think" with GIS.

The question of thinking with GIS moves well beyond the case-study examples of how someone else used the system to produce a land-suitability assessment, or to establish habitat regions to protect environmentally sensitive lands. Ample case studies can be found in the GIS and environmental planning literature alike; what is uncommon is discussion detailing how those results were derived (i.e., which tools were used with what data and how). When looking only at the results, GIS remains an abstraction, a "black-box" whose inner workings are just plain mysterious. Of course, nothing could be further from the truth. When one is armed with this "nuts and bolts" information, GIS can be more fully appreciated and its ability to support environmental planning better exploited.

The value of GIS as a tool for developing planning support systems is best assessed with reference to the nature of the scientific input required at various stages in the decision-making process (Webster, 1993). A useful model that identifies where GIS fits, underscoring its potentials and limitations as a planning tool, has been introduced by Webster (1993). According to this framework, thinking with GIS begins with the information needs encountered during each stage of the planning process. From problem identification, goal-setting, evaluation of alternatives, selection, implementation, and monitoring, specific scientific support needs can be expressed.

Problem identification

This initial stage in planning can be understood to involve the measurement of demand for public goods that requires the measurement of negative externalities (Webster, 1993). Measurement can focus either on the identification of current problems or the anticipation of future problems. Two forms of scientific support are essential during this stage: description, followed by prediction. Descriptive analysis draws upon spatial data as a means to display patterns of demand together with existing patterns of key public goods. Expressing patterns implies the ability to ask questions of the data and visualize the results related to landscape features such as the location and characteristics of:

- Existing infrastructure.
- Existing demand points.
- Discontinuities related to poor allocation.
- The spatial expression of recently implemented policies.

Prediction concerns questions pertaining to the quantity, structure, and location of consuming units at some future point in time. This type of forecasting requires both a store of data and the ability to organize and model that data to produce planning scenarios that can be explored via the GIS.

Goal-setting

Setting goals can be thought of as the attachment of weights to different market or social preferences. The support required to rate, rank, and weigh preferences is highly prescriptive, with GIS analysis directed toward the aggregation of individual and group attractiveness scores using some form of implied social-welfare function (Webster, 1993).

Plan generation

In addition to the need to sample various solutions and alternative interpretations, there is a need to examine the probable geographic pattern created by planning options and study their ramifications. In this context, a given solution or recommendation expressed in the plan can be viewed as a hypothesis, where testing the hypothesis requires support to search the solution space and determine the feasibility of a given alternative. Sampling the solution space is accomplished through a form of prescriptive analysis easily conducted using GIS.

Evaluation and selection

This planning phase describes the process of narrowing down a set of feasible alternatives to a single "optimal" solution. The process, while guided by expert judgment, is another example of GIS-based prescriptive analysis where weighted goals are examined and placed into a "model" that follows some form of optimization logic. The rating and weighting process used to explain landscape variables for suitability assessments is a classic example of this procedure. In general, optimization identifies a series of procedures that will either maximize some potential surface related to the weighted goals or minimize a constraint surface.

Implementation

Implementation is a management task where specific recommendations are placed into action. Here, a series of processes and outcomes follow from a set of specific directives that require descriptive, predictive, and prescriptive support to track progress, adjust strategies, and provide feedback to policy-makers. The map that accompanies the comprehensive or environmental plan and shows where things should be, or how the planning area should evolve, is an example of this type of product.

Monitoring

Keeping pace with the driving forces that continue to shape the planning area requires the "on-going" sensing to the landscape in order to identify new problems and trends and to track the progression of the plan. Key scientific support given by GIS includes descriptive analysis and prediction. Descriptive support is required to address fundamental "where," "what," and "how many" types of questions that will recover useful information from the database on the current status of selected indicator variables. Predictive support is needed to extrapolate trends and explore the consequence of unfolding time-dependent patterns and processes.

GIS solutions build from these scientific inputs and the types of answers required in order to satisfy basic planning questions. From this point on, analysis is simply a matter of connecting the functional capabilities of the GIS to a descriptive, prescriptive, or predictive methodology. Obtaining planning solutions using GIS therefore requires a problem-solving strategy or script. Two common strategies can be followed to transform data into the information that satisfies a planning need: a top-down strategy or a bottom-up strategy (Eastman et al., 1993). With either approach the goal is to simplify analysis and reduce uncertainty to yield a useful information product.

Top-down analysis

The top-down method of GIS analysis develops from a clear and well-defined problem. Following this strategy, the problem under investigation is divided into smaller and more manageable parts. Concentrating on each individual element of the problem, each subsolution can be addressed and then their results combined at the appropriate time to produce the analytic solution to the larger question. Scripting the analysis in this manner suggests that for each subproblem a submodel is constructed whose result is an intermediate step to the larger problem. Therefore, the final expression develops from the logical combination of each submodel. To illustrate the use of this strate-

gy consider the problem of developing an environmentally responsive zoning designation for a newly incorporated area of land. This prescriptive allocation problem can be divided into a series of subproblems based on a set of individual land uses whose suitability can be assessed with respect to the environmental constraints imposed by the site. Using GIS, important landscape variables can be examined together with the patterns of present and future infrastructure characteristic of the area. Then, on a land-use by land-use basis constraints can be identified and a potential surface can be generated to show where optimal conditions exist. Combining each of these separate models together using logical overlay techniques creates a composite profile for the area that explains a possible zoning plan. Within each zone, more detailed studies can be performed that can help determine specific developmental densities, design constraints, and other factors that can be translated into specific zoning regulations. In this example, by working on each of the smaller aspects of the problem, analysis is greatly simplified and the factors that control or influence the problem can be more easily isolated. In addition the analyst enjoys greater control over the combination of factors, their classification, and how they can be assembled to generate the solution.

Bottom-up analysis

The bottom-up strategy reflects the converse of the top-down approach. Bottom-up analysis is employed in situations where GIS is being applied in an exploratory of experimental manner. Using this approach the decision-maker begins with the smaller components of the problem or hypothesis that is being tested. Because the "big picture" may not be well understood, very little concern is given to a detailed definition of the problem. Thus, because the analyst may not know what the solution should look like, working with smaller and better-understood relationships allows a possible solution to emerge through their analytical combination. Using the GIS, facets of the problem are examined and assembled into successively larger and more complete answers based on either theory or intuition. Therefore, the GIS model builds upward toward a solution through trial and error logic. Although the bottom-up method may not be a reliable problem-solving strategy, it can prove to be useful in situations where evidence of the larger problem is uncertain or where guiding theory is absent. A good example of an environmental problem that can be approached using a bottom-up design is the question of cumulative environmental impact assessment. With respect to cumulative impact analysis the question as to how individual projects with minimal environmental significance can introduce cumulative effects that form serious adverse impacts lacks a clear and understood theory. However, taking these individual projects and aggregating their spatial pattern via GIS can gain insight into the larger question of cumulative change. Here, each smaller pattern or component builds upward to create a more comprehensive definition of the problem. Using "bottom-up" analysis, decision-makers can identify the causal processes at work and describe the geographic factors that contribute to the "big picture" of cumulative environmental change.

The decision problems to which these organizing strategies may be applied fall into four main analytical categories:

1 Single-criterion single-objective problems – here we may be dealing with the situation where one variable, such as slope, can be used to determine whether a site can support a specific type of land use.
2 Single-criterion multiobjective problems – in this case we might be able to use a factor such as slope to help illuminate several issues such as run-off, slope stability, erosion potential.
3 Multicriteria single-objective problems – these situations explain the multidimensional nature of some problems, where to form a solution it would be necessary to assemble all the variables that impart some influence. Examples might be the use of soil type, aspect, temperature, land use to explain the feasibility of reforestation efforts.
4 Multicriteria multiobjective problems – with this type of problem we find a set of controlling variables that influence or illuminate a range of problems, such as the use of demographic data to help understand key characteristics of a neighborhood.

These four classes of analysis identify the range to which GIS data may be applied and point to the methods needed to arrive at a solution. The specific GIS solution takes form as a decision rule that explains the procedures followed (required) to guide the combination of criteria taken from the database and the manner by which these data will be treated using the GIS. Thus, a decision rule will involve a particular set of operations that will be performed on the data. These operations may include setting a threshold value during a search and identifying a specific attribute, or a more complex sequence of operations that will transform, manipulate, combine, or compare data.

To the analyst/planner, decision rules become the directions that explain how a task will be conducted and which operations are needed to produce a solution. These rules or directions can be created for a variety of reasons, such as:

- The manipulation of attributes to reduce information content by grouping, isolation, classification, scaling procedures.
- The manipulation of attributes to increase the information content by ranking, evaluation, and rescaling functions.
- The combination of pairs of input values through cross-tabulation, correlation testing, principal components analysis.
- The logical combination of data through map overlay based on contributory rules, dominance rules, or interaction rules.
- The application of distance operations through buffering operations.
- The generation of surfaces through the application of neighborhood functions, statistical models, or interpolation methods.

A central feature when developing a decision rule concerns the methods used for selecting a means of "deciding." Deciding considers the overall approach that will be followed to reach a solution to the problems, and requires treatment of how the relevant data will be selected, how that selected data should be categorized, and how the data are to be logically combined. The methods for deciding can either rely on the formulation of a choice function, or a choice heuristic (Eastman et al., 1993). Choice functions are a mathematical means of comparing alternatives that employ some form of optimization that blends linear programming techniques with GIS data-processing functions. Choice heuristics define specific procedures that will be followed using GIS functionality exclusively. Because heuristics rely on tools provided within the given GIS they tend to reflect the more typical framework of GIS supported decision-making.

Formulating decision rules based on choice heuristics follows the general methodology referred to as map (cartographic) modeling (Tomlin,

1990; Berry, 1995) or map analysis (Bonham-Carter, 1994). This technique is founded upon the use of a map algebra that consists of a sequence of natural language commands or concepts that perform key data manipulation functions. With these commands mapped, data can be analyzed and a series of operations can be articulated that will create the desired derivative product. Samples of the map modeling commands common to GIS are listed in Table 8.3. With these commands, statements can be created that when followed produce a map model that depicts a geographic arrangement or relationship required by the analysis. To implement these commands effectively it is generally good practice to express the sequence of modeling operations in the form of a flow diagram

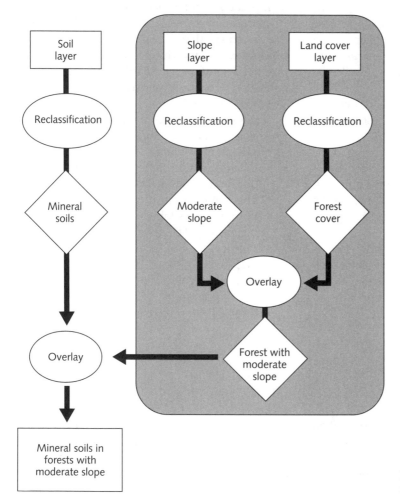

Fig. 8.7 A generalized cartographic modeling example.

(Fig. 8.7). Using this simple device, the problem can be decomposed into a number of analytical steps with the data and GIS operation specific to generating that step clearly represented. Reviewing the flow of analysis from these diagrams gives the analysis an opportunity to spot logical flaws, missing steps, or poor problem definitions prior to using the GIS. Above all, the exercise reinforces the geographic "thinking" that must precede any GIS application.

The question of error

As GIS methodologies find acceptance and wider application within the environmental planning community, the issues of data quality and the sources of error that can adversely affect the results of GIS analysis must temper their use in decision support. In the drive to solve problems, the question of spatial data quality can be easily overlooked, yet no map stored in a GIS is completely error free. As a consequence any analysis undertaken using GIS will not only provide useful information to guide planning, it will also contain error. The sources and treatment of error greatly influence the correctness of inferences drawn using the results produced from the GIS.

The possible sources of error in a GIS are numerous. Detailed discussions identifying the sources and treatment of error in a GIS can be found in Muller (1987), Veregin (1989), and Burrough and McDonnell (1998). When considering the quality of data used to address a planning problem it is important to remember that the data contained in the database has its origins in the planning area. From the field to the database the observations recorded have been measured, classified, interpreted, estimated, and encoded by many hands and under contrasting conditions. As a result it can be expected that this data contains errors to some degree. However, even after the data has been entered into the GIS, error can propagate and accumulate as it is manipulated and analyzed by the user. Although some users of GIS may be aware of error and how it propagates, in practice little attention is given to its significance.

Presently, much of the work related to error and error propagation in GIS remains at the research level. Nevertheless, the factors affecting the quality, reliability, and assumptions inherent to spatial data have been identified (Burrough & McDonnell, 1998). As suggested in Table 8.8, GIS analysis is no different from any other analytic technique, and the same caveats apply to GIS as they do with respect to statistical or numerical modeling. Error simply takes different forms. To illustrate this point consider the stages followed when developing the GIS database. When spatial data are entered into the GIS it is often accomplished by means of digitizing a map that had been originally drafted on some type of sable polyester film media. That map, regardless of the effort that went into its production, contains uncertainties that are duplicated in the GIS. As many planners, geographers, or other field scientists know from experience, carefully drawn boundaries and contour lines on maps are elegant

Table 8.8 Sources of error in GIS.

Accuracy of content:
- density of observations
- positional accuracy
- attribute accuracy
- topological accuracy
- linage

Data measurement:
- data entry faults
- data model generalization
- natural variation in the data
- observer bias
- processing error

Analysis and modeling:
- classification and generalization problems
- map overlay problems
- choice of analytic model
- flawed logic
- error propagation effects
- interpolation error

Reliability factors:
- age of the data
- areal coverage
- map scale/resolution
- relevance
- format

misrepresentations of changes that are often gradual, vague or fuzzy (Burrough & Frank, 1996).

The accuracy of mapped data placed into a GIS is typically expressed in three ways: (1) thematic accuracy which describes the correctness of the attributes characterizing the map theme, (2) positional accuracy which defines its precision with respect to some locational coordinate system, and (3) temporal accuracy explaining how representative that map and its patterns are as a function of time-dependent processes. Yet error can occur anywhere in the process from observation to presentation, including conditions such as:

- Errors in perception
- Errors in approximation
- Errors in measurement.

In addition to these listed, error can be introduced when data is stored in the computer because too little computer space was allocated to store the high-precision numbers needed to record data at a given level of accuracy. Then there are the situations where some data may be too expensive or too difficult to collect and inexact correlates are used with cheaper-to-measure attributes (Burrough & McDonnell, 1998). These realities of spatial data usually lead to the opinion that error and uncertainty is bad. Obviously it would be better if error did not exist, but that position is unrealistic. To the analyst applying GIS to address a problem, it can be extremely valuable to know how and where error and uncertainty occurs and to explore methods to manage and reduce its impact. By so doing, we can improve our understanding of geographic processes and the patterns they form.

One means of developing confidence in the data component of a GIS is through the implementation of a formal error assessment procedure (Dunn, Harrison, & White, 1990; Finn, 1993). Standard error assessment techniques are conducted by comparing attribute values in the database with values from known ground locations. This process of determining the "ground truth" is comparatively straightforward and can provide the decision-maker with an expression of error that effectively characterizes the thematic accuracy of the digital file (Lein, 1997). Positional accuracy, a more complex variable to measure, can be examined through the use of GPS samples taken systematically at selected ground control points and along major geographic features that trend through the area. These measurements can be used to verify and correct the geometry of the digital file.

The presence and propagation of error influence how the planner will utilize the support functions offered by the GIS. Thus, despite the promise of this technology, GIS is not a panacea (Aageenbrug, 1991). By adopting a more considered view of this mode of spatial data handling and analysis, the applicability of GIS and its limitations can be grounded in empirical evidence. When this occurs the benefits ascribed to GIS can be realized:

- Improved quality of information
- Improved timeliness of information
- Improved information flow
- Improved efficiency
- Improved productivity
- Improved decision-making.

Managing GIS projects

In a public-policy setting the potential uses of GIS are numerous. Worral (1991) lists several that are particularly relevant to the environmental planner:

- The more sensitive monitoring of change in demographic, social, economic, ecological, and environmental conditions.
- Developing a better understanding of the processes of change and the complex interactions between components of the region.
- The more accurate forecasting of the changing needs for publicly provided services.
- The more precise identification of spatial variations in living conditions as a basis for the development of social policy and the more precise targeting of local government resources.
- The more rigorous identification of target markets for the promotion of local services.
- More effective and responsive service planning by more accurately identifying the determinants of demand and by more expertly forecasting the changing pattern of need for

services as a basis for setting priorities in the deployment of resources.

- Improving statutory planning processes by developing the means for modeling and simulating alternative scenarios, and by developing the techniques to assess the suitability or conformance of developmental proposals in the context of statutory plans.
- Improving the policy-making process by developing more sensitive methods for the evaluation and analysis of policies and programs.

All of the above are possible provided GIS technology can be brought into local government and the entities involved can resolve the issues associated with justifying and managing GIS projects.

The management issues surrounding GIS have been discussed in detail by Huxhold (1991), and Huxhold and Levinsohn (1995). Three central themes have been identified that direct GIS management:

1 Evaluating geographic information needs – this management focus considers the importance of developing long-range GIS plans that are consistent with the goals of the organization and determining the geographic information needed to achieve those goals.
2 Maintaining organizational support – directs management to engage in cost/benefit studies to determine the economic value of GIS, and to conduct pilot projects to test various GIS design and application possibilities.
3 Managing the GIS project – considers the management problem associated with converting all mapped information to digital form and organizing resources for continued support and future expansion.

These management themes suggest that GIS is more that a software package installed onto a computer. Management concerns also imply that when placed into a policy-making setting GIS is also more than a problem-solving tool. Rather, placing GIS into the planning process institutionalizes the technology so that data, computer hardware and software, people, and procedures are available and dependable when they are needed (Huxhold & Levinsohn, 1995).

Perhaps the most critical aspect of the GIS management challenge is the development of a long-range GIS plan. This plan must relate to the long-range goals of the local government it is designed to serve. A comprehensive GIS plan analyzes the needs of the organization over a 5- to 10-year time horizon and attempts to identify how GIS will be used and avoid unrealized expectations and disappointment. For the decision-maker the plan allows for the careful evaluation of the system's applicability to the strategic and tactical plans initiated by local government. The plan also ensures that the appropriate resources will be available where and when they are required (Huxhold, 1991). To provide this capacity the long-range plan must address five critical programmatic needs:

1 Obtaining high-level support – not only does the long-range plan bring opportunities for improving government performance to the attention of decision-makers, but it also gives them confidence that those who advocate the new GIS technology are competent and can make the project succeed.
2 Identifying all potential application – a comprehensive long-range plan considers all of the geographic information needs of the organization and thus ensures that no improvement opportunities are omitted from consideration.
3 Prioritize applications for orderly implementation – if a long-range plan is successful at identifying all potential applications and is related to the strategic and tactical plans of local government, then it is possible to schedule or prioritize the implementation of applications in the order that will be most beneficial to the organization.
4 Obtain maximum benefits organization-wide – since one of the major features of a GIS is its ability to integrate information from a number of different sources, the long-range plan can identify information-sharing opportunities that had never before been realized as possible without a GIS. Separate functions that had never shared information before can realize improvements through geographic data integration.

5 Identifying resource requirements – one of the worst mistakes information specialists can make when advocating GIS is to surprise decision-makers later, after the project is well underway, with the need for additional unanticipated funds. Since decision-makers may not be familiar with GIS technology they place considerable trust in their technical staff to present a comprehensive and understandable proposal upon which funding decisions can be based. When costs escalate, the credibility of those in charge can be compromised.

A number of information-system planning methodologies have been introduced (Huxhold, 1991). These methodologies are designed to ensure that new technology will successfully support the long-term goals of the organization. Of the methods introduced, three of the more relevant to the implementation of GIS are: (1) business systems planning, (2) critical success factors, and (3) the Nolan–Norton stages theory. Each method listed above seeks to define the functions and other factors that are most important to an organization and then develop the appropriate information system strategy that offers the most beneficial support.

The most widely accepted model of computing growth in an organizational setting was advanced and refined by Nolan (1973; 1979). The model is described in detail by Huxhold and Levinsohn (1995) and serves as the GIS management centerpiece of the strategic planning process. The Nolan model uses a six-stage process to explain the growth and evaluation of computing resources in an organization.

1 Initiation – the initial stage where external technology advancements become known by users and applied to functions requiring cost containment.
2 Contagion – explaining the case where growth in applications increases significantly and users become more skilled in the technology.
3 Control – the stage of continued growth causes rising costs which attracts managements attention, and formal planning and controls are implemented.

4 Integration – as the system matures, management identifies the technology as an organization-wide resource and applications are modified to take advantage of database management capabilities.
5 Data administration – giving emphasis to shared data and systems where applications become widely integrated as a total organizational resource.
6 Maturity – the culminating stage where applications are designed to follow the flow of information through the organization, and computing technology is accepted and used as a strategic resource by users as well as managers.

Accordingly, the Nolan model assumes that the new technology enters an organization and is adopted at the user level where it becomes popular. However, different planning agencies and local governments are most likely to be in different Nolan stages; consequently, they are also likely to progress through these stages at different rates. This observation helps to explain why the adoption of GIS technology has not been more widespread than expected (Huxhold & Levinsohn, 1995).

Managing the work required during the operational phase of a GIS consists of defining projects, developing a work plan to apply resources to the projects and schedule their completion, and preparing a budget to make certain that the financial, machine, and human resources will be available to complete the project. A project can be defined as any work activity that requires GIS resources to produce an end-product of value. In general, GIS projects can be categorized as direct or indirect. Direct projects have a beginning and an end with specific time limits attached and understood. Indirect projects are less well defined, often last an entire work-year, or are very repetitive. Based on the nature of a project, a work plan will be needed to determine the resources that it will require and to facilitate accurate budgeting. Several factors are important to this work plan, including:

• An estimate of how many staff hours will be required to complete the project.
• An estimate of the total cost of the project

including staff time, software, hardware, and other relevant costs.

- An estimate of when the work on the project will begin and when it is scheduled to be completed.

This information is critical to the success of the GIS project management system and to effective budgeting for GIS operations. The information provided in the work plan also forces the planner to realize that GIS planning support exists within the constraints of the capital budgeting programs of local government, Thus, while GIS has demonstrated potential, there can be a gap between what is possible and what is practical when GIS is placed into day-to-day operation. Effective project planning together with a more considered view of GIS technology helps to narrow that gap.

Beyond conventional GIS

At the outset of this chapter, GIS was characterized as a dynamic and adaptive problem-solving methodology. However, the question remains as to whether GIS has been applied to its full potential, or if it alone is sufficient to provide the technical support required by the planner (Holmberg, 1994). In response to this question the concept has emerged of a planning support system that employs the best available GIS technology as its technical core, and then branches beyond GIS to incorporate IT environments that provide more specialized forms of a planning assistance. Recently, there have been several examples where GIS has been reconfigured to serve as the nucleus of a planning support system (Faust, 1995; Zhu, 1998; Shiffer, 1995; White et al., 1997; Batty, 1997). These extensions beyond conventional GIS grow out of a new dynamic between science and technology that encourages a change in the manner by which environmental information is managed and analyzed (Stafford et al., 1994). At issue is the observation that the design, implementation, and operation of information systems to support planning are inflexible. Many of the deficiencies surrounding GIS have been attributed to defects in handling, representing, visualizing, and commu-

nicating diverse forms of geographic data. Overcoming these support deficiencies has contributed to the design of hybrid systems based on advances in multimedia/hypermedia systems, knowledge-based systems, virtual environments, and web-based technologies. Although still very much in the development state, these emergent technologies are likely to redefine the concept of GIS and the reshape the future of computer-assisted planning.

Multimedia extensions

The term multimedia has been defined as the communication of information in a variety of formats including text, graphics, images, animations, video, and audio. Multimedia systems promote the use of integrated media in a highly interactive format that (1) makes digital information more directly accessible and (2) places the user of digital information in a more active role. Thus, the incorporation of multimedia capabilities in a planning support system encourages access to new data sources (images, video, sound), introduces enhanced visualization and manipulation capabilities, and fosters creation of more intuitive multimedia interfaces.

The main characteristics of multimedia systems that contribute to the design of hybrid GIS are those that influence the assimilation and presentation of information to the user:

- Nonlinear data access – multimedia designs are highly interactive and based on the principle of permitting users to move through data in a nonlinear way.
- Interactivity – interactivity is the characteristic principle of multimedia technology. Interactivity is achieved by means of hypertext where words or images can be connected with other information that can be accessed by use of pointing devices. This technique permits a very simple and immediate ability to navigate through data.
- Ease of navigation – navigation describes the user's movement through the multimedia application by means of the interactive instruments employed in the systems design. The greater the range of interactive instru-

ments, the greater and more flexible the navigation. Ideally, navigation implies that movement through the system is not rigidly structured, but as random and possible.

- Substituting speech for text – text often becomes overwhelming and difficult to comprehend fully on the computer screen. Integrating maps and images with speech can give detailed descriptions that may allow users to concentrate more on the visual interpretation of data.
- Use of animation – animation consists of giving movement to a scene or a figure on a fixed background. Such a technique can permit easier consultation of thematic maps and aerial images where animation can be used to highlight important information that might make complex phenomena easier to understand and visualize.
- Use of morphing – morphing is the method used to transform an object into another by means of animation. This technique can be instrumental in representing dynamic events and give planners a way to vary the descriptive characteristics of an object over a continuous range.

Intelligent systems and applied AI

Since the early development of computers there has been a strong interest in finding ways to make these machines smarter. Over the past decade the prospects of designing intelligent information systems based on Artificial Intelligence concepts have been explored (Lein, 1997). AI technologies, such as the expert system, can contribute to the development of decision tools that can

- Learn or understand from experience.
- Interpret ambiguous or contradictory information.
- Use reasoning strategies to solve problems.
- Respond to new situations.
- Deal with complex situations.
- Acquire and apply knowledge.

A detailed literature examining the use of expert systems in environmental planning has emerged (Hushon, 1990; Kim et al., 1990, Wright et al., 1993). Application areas that show promise for the design of intelligent systems include environmental impact analysis, developmental suitability mapping, and zoning (Lein, 1989; Geraghty, 1992; Han & Kim, 1989). However, the most promising work to date has been in the development of knowledge-based spatial decision support systems (Kirkby, 1996; Zhu, Healey, & Aspinall, 1998).

Virtual planning environments

The prospects for developing computer systems that can produce representations of a problem in the form of a mental context-model that can be experienced within the machine environment are both tremendous and challenging (Raper et al., 1998; Faust, 1995). Although true virtual geographic or planning support systems have yet to be implemented, several requirements have been identified that will be instrumental in creating true virtual systems:

1 vicarious travel
2 multimedia
3 three-dimensional modeling
4 virtual reality.

Developed with these capabilities, an interactive three-dimensional virtual reality GIS would have to deliver a very realistic representation of the planning area, provide a means to permit free movement within the selected geographic area, perform all normal GIS functions employing a three-dimensional database, and view the results from any vantage point, integrating visibility functions into the user interface. To date, creating virtual environments has proved to be difficult primarily because the data sets from which the applications are created must be collected firsthand and the software environments must be developed from scratch for each application.

Summary

In this chapter the computer and its realization as a form of planning support was examined. Key concepts from the field of decision support were explored in this context, and the role of the computer as a decision support environment was described. Decision support was discussed with an emphasis on the prospects for automating key decision problems and facilitating problem structuring when issues are ill-defined and require novel solutions. The nature of geographic information systems was a central theme in this discussion. Because GIS has become an essential part of the planner's toolbox, its design and implementation, together with its role in data analysis and communication, are key elements in supporting planning decisions. Yet GIS is not a panacea and the concerns surrounding error and its propagation where noted. The chapter concluded with a review of newly emerging information technologies that offer new capabilities to represent and study environmental planning problems.

Focusing questions

Discuss the concept of decision support with reference to planning problems that are ill-defined.

How can an understanding of the error inherent to GIS analysis influence a more considered use of this technology?

Where can decision support be applied in environmental planning?

Discuss the factors that influence GIS implementation.

How does GIS change how environmental planning is conducted?

CHAPTER 9

Ethics, Conflict, and Environmental Planning

Environmental planning has been characterized as a decision-making process where the planner attempts to achieve a sustainable balance between human needs and environmental protection. This planning specialization places a high priority on environmental matters concerning land use, policy, and design, and as a decision process, encourages decisions that maximize benefits to both people and the environment in which they reside. Yet, any decision-making process is confronted with the larger issues of not simply making decisions, but with the more daunting responsibility of having made the correct decisions. Recognizing that making decisions is often very different from justifying those decisions, there is more looming behind the environmental planning process than a plan or a decision, there is a landscape of beliefs that injects its influence and colors our perceptions and approaches in obvious and subtle ways (Hargrove, 1985; 1989). There are no easy answers to the issues that surround an environmental planning decision, however there may be a set of principles that can be called upon to aid and guide the planner when decisions must be made. In this chapter we will examine the emergence of ethical standards that can direct planning practice and explore the question of ethics in planning and the value of applying these "rules" to help guide our environmental judgment.

Ethics and choices

Whether we realize it or not, each of us possesses a set of values and calls upon those values as we make judgments and decisions in our daily lives. In the practice of planning those values often introduce themselves in ways we may not be aware of, yet their influence can be undeniable. Situations we encounter constantly ask us to make choices, and when we choose we are expressing our values, rating things as better or worse, important or unimportant, good or bad. While some choices are trivial and carry no significant consequence, others can be monumental, affecting conditions in the planning area for years to come. Because choice is inevitable, the question is not whether we shall have preferences, standards, or ideals; the real question is whether our choices will be consistent or inconsistent, life enhancing or life destroying (Percesepe, 1995). Although there is a tendency to pretend that environmental planning is "value-free" and purely objective, we have come to recognize that value judgments exist and play an active role in shaping environmental decisions (Lemons, 1987). Ignoring the role of value judgments in the practice of planning leads to a lack of wholeness and perspective. Therefore, if we are to choose well, we need to acknowledge the personal and social values we bring to the planning problem. Acknowledging the presence of value judgments and considering the choices we make that draw upon them directs us to the subject of ethics and its influence in the planning process.

Ethics may be defined as the branch of philosophy that seeks to systematically analyze moral concepts and to justify moral principles and theories (Percesepe, 1995; MacKinnon, 1995). Therefore, ethics, or moral philosophy, asks us to consider foundational questions about "good," "bad"; about what is "better" and "worse," about whether there is an objective "right" or "wrong" and how we can know if there is. Through the study of ethics we endeavor to establish principles of right conduct that can serve as a guide to our decisions, and in its search for those values ethics constructs and criticizes arguments that state valid moral principles and the relationship between these principles (Percesepe, 1995). Within this definition, it is assumed that the primary objective is to help us decide what is good or bad, better or worse, either in some general way or with respect to particular ethical issues (MacKinnon, 1995). This focus is referred to as normative ethics, and when it is compared to the natural sciences, which are largely descriptive and concerned with empirical facts, ethics tends to be far more prescriptive and directed toward an understanding of normative values. In this context, we can describe ethics as a normative-based study concerned with the discovery of "what ought to be" rather than "that which is" (Percesepe, 1995).

To the environmental planner ethics becomes something that is practical, action oriented, and seeking to affect change from what is to what should be. Although ethics does ask very general questions about the nature of "good" and "bad," its value to the environmental planner rests in its aim to help us determine the right or better thing to do in particular situations. For example, when faced with the question of saving jobs or saving an environmental amenity, or when confronted with a situation that is known to hold significant environmental risks, ethics serves as the foundation that supports those decisions that have to be made, and provides guidance on what should be done. The central feature of the normative ethic called upon in the situations we confront is the moral principle. Moral principles are general

guidelines for right conduct and possess certain distinguishing characteristics that are worth noting (Pojman, 1990):

1 Prescriptivitiy – a term that refers to the action-guiding nature of morality. Generally, moral principles are posed as injunctions or imperatives.

2 Universalizability – a concept best exemplified in the notion of a "Golden Rule" that prescribes what is right for one person is also right for another in similar situations. Universalizability is the root requirement of impartiality such that when formulated as a principle it functions as a rule that forbids us from treating one person differently from another.

3 Overridingness – suggests that moral principles take precedence over other kinds of considerations including aesthetic, prudential, and legal ones. Thus legal justice and moral justice do not necessarily coincide, since laws themselves may be morally unjust. Legal justice depends not only on whether there is law, but also on the overriding question of whether law recognizes and protects the moral rights of those affected.

4 Publicity – recognizing that we use principles to play an action-guiding role in our lives. For those principles to be maximally effective they must be made public.

5 Practicability – moral systems must be workable and their rules must not lay an excessive burden on moral agents. Overly idealistic principles may produce ineffective action if human limitations are not taken into consideration.

These root characteristics point to the fundamental traits that ethical principles share in common. Equally important is the requirement that one must have reasons to justify one's moral conclusions. This does not suggest that making ethical judgments is a purely rational activity. As MacKinnon (1995) notes, we might be tempted to think that in order to make good moral judgments, we must be objective and not let our emotions enter into our decision-making. At times this position may be valid, particularly when anger, fear, or jealousy bias or prejudice our thinking. Yet emotion need not be divorced from decision-making, provided we can explain why we hold a certain moral position. Simply stating that "*x* is wrong" is generally not sufficient (MacKinnon, 1995). To consider this point in more detail we need to recognize that ethical statements and judgments are evaluative. When voiced they tell us what the individual believes is good or bad and they further express a positive or negative regarding the object of their judgment. Although factual information is relevant to our moral evaluations, we may not "see" that part of the argument until the moral conclusion is justified.

Ethical judgments rely on empirical and experientially based information. When interpreting ethical judgments it is useful to distinguish between those that are empirical or descriptive and those that are evaluative or normative. Descriptive judgments are those in which we state specific factual beliefs. Most moral judgments tend to be evaluative, for they "place a value" on some action or practice. Because these evaluations also rely on beliefs in general about what is good and right based on norms or standards, they are also normative. In making ethical judgments our decisions are often based on the use of terms such as good, bad, right, wrong, obligatory, permissible. When these terms are used our interest is focused on what we should or should not do, making these terms function in a purely evaluative context. However, when we speak of a good land-use policy we are probably not attributing moral goodness to the plan. Therefore, it is equally important to realize that not all evaluations are moral in nature. With respect to our land-use example, our reference may be to practical usefulness and efficiency, even though the policy may have moral implications. Consequently, we can distinguish the various types of normative judgments and the areas where such judgments are made from descriptive judgments about factual matters and areas that are exclusively descriptive (MacKinnon, 1995).

Planning and relativism

Because the planner does not operate in a vacuum, not only is it important to understand the basic constructs of ethics, but also to form an awareness that while ethics may strive to identify universal principles, reality may suggest that the landscape in which planning takes place may be more aptly characterized by a type of ethical relativism. Ethical relativism is rooted in the doctrine that the moral rightness and wrongness of an action will vary from society to society. Based on this idea, relativism suggests that there are no absolute universal moral standards binding all people at all times (Ladd, 1973). Because ethical values and beliefs are considered relative to the various societies that hold them, there is no objective right or wrong. The ethics or standards held by a group are a function of what those societies believe. From an environmental perspective, relativism may provide useful insight into much of the debate surrounding environmental protection, land degradation, and the social basis for planning.

We can better understand ethical relativism by comparing our views of ethics and ethical matters with our beliefs about science and morality (MacKinnon, 1995). Most people envision the natural sciences as objective and governed by a generally accepted method that has produced a common body of knowledge about the natural world. Morality does not appear nearly as objective. Furthermore, there is no general agreement about what is right or wrong and there is constant doubt about what and how we agree. As a result, there is no objective moral truth or reality comparable to that which we seem to find in the natural world. Thus, we tend to consider morality as a matter of subjective opinion that supports the basic conclusion of ethical relativism.

There are two expressions of ethical relativism that have important implications in environmental planning. The first is personal ethical relativism, where judgments and beliefs are simply the moral outlook and attitude of the individual. Using this idea, we can consider the planning area as a landscape comprised of individual beliefs based on personal histories that define why an individual holds certain views or attitudes. Therefore, we can assume that there will always be conflicting perspectives – those who like trees and those who see nothing wrong with cutting them down, those who like growth and those opposed to any form of development at all. While we may attempt to educate or persuade people to change their beliefs, declaring either position right or wrong is difficult, since we must assume an objective standard against which the correctness of their beliefs may be judged. A second version of ethical relativism has a social or cultural context. This form of ethical relativism holds that ethical values vary from society to society and that the basis for moral judgment lies in those social or cultural views (MacKinnon, 1995). Individual right or wrong is viewed using the lens of social norms.

Arguments supporting the theory of ethical relativism are based on beliefs surrounding the diversity of moral views, the presence of moral uncertainty, and complexities introduced by situational differences. However, relativism if taken to the extreme can introduce several undesirable consequences (Rachels, 1998). First we could no longer say that the customs or practices of other cultures are morally inferior to our own. Secondly, we could decide whether actions are right or wrong simply by consulting the standards of our own society. Lastly, relativism calls the idea of moral progress into doubt.

Obviously, the concept of moral relativism is not easily digested. The purpose of introducing ethics and relativism in environmental planning is to underscore the fact that morality is more complicated than simply a matter of right or wrong. By considering ethics and its variants we can improve our ability to comprehend the ethical judgments we face and the difficulty we encounter when ethics are placed into an environmental context.

Toward an environmental planning ethic

To begin our discussion of environmental ethics we can examine an argument advanced by Regan (1995) dealing with the nature and possibility of an environmental ethic. Regan posed the question as to whether an environmental ethic is, in fact, possible given that to have an environmental ethic presupposes that we agree on the nature of what

this ethic must be like. Since universal agreement on the topic of what ethics apply to the environment has yet to occur, the conditions such ethics must meet can only be suggested. According to Regan (1995), an environmental ethic must be grounded on the basic premises that:

1 Nonhuman beings have moral standing.
2 The class of those beings that have moral standing includes, but is much larger than, the class of conscious beings.

If both conditions are accepted, then a theory that satisfies neither of them is not a false environmental ethic; it is not an environmental ethic at all. Should we continue with this line of reasoning, then any theory that meets our first condition but does not satisfy the second may be considered as a theory on the way to becoming an environmental ethic; but since it fails to satisfy the second condition, it fails to qualify as a genuine environmental ethic.

The significance of these two conditions is that they assist us in distinguishing between an ethic of the environment, and an ethic for the *use* of the environment. In environmental philosophy this distinction is embodied in the contrast between kinship theories and management theories and the standards they impose. Kinship theories have developed out of the idea that beings resembling humans to the extent that they are conscious, have moral standing; whereas management theories direct us to preserve or protect the environment if it is in the interest of human beings. When either theory is applied, the possibility that human and animal interests might conflict with the survival of nonconscious beings becomes possible. As this occurs, it is doubtful whether such conflicts admit to rational adjudication (Regan, 1995).

Reconciling the question of ethics and environment requires conceptual strategies that reduce logical conflict. One general strategy is that of assimilation. Assimilation is the ideal that existing moral categories and conceptions of value are adequate for describing our environmental concerns. While attractive because of its simplicity, assimilation fails when confronted with new cases that do not fit existing concepts. Assimilation ethics, therefore, form out of (1) the application of existing values to a re-describe concern about human beings and our relation to the environment or (2) the extension of concepts such as good, right, and value to nonhuman beings. A second strategy is one of challenge rather than assimilation. Thus, instead of finding values in rocky desert landscapes, tropical forests, or ecosystems based on anthropocentric rationales, existing values are challenged and the argument between the possession of rights and the possession of interests is replaced with the ethic that natural things, although lacking "interest," remain worthy of respect (Brennan, 1995).

Based on the general positions advanced above, an environmental ethic becomes a working hypothesis that links humans, nature, and values to include:

- A theory about what nature is and what kinds of objects and processes it contains.
- A theory about human beings, providing some overall perspective on human life, the context in which it is lived, and the problems it faces.
- A theory of value and an account of the evaluation of human action with reference to nature.
- A theory of method, indicating by what standards the claims made within the overall theory are to be tested, confirmed, or rejected.

As we consider the types of decisions made in planning and the balance that must be achieved between human need and environmental functioning, it might be helpful to put the concept of an environmental ethic in perspective and ask whether such an ethic is really needed. First, in addressing this question it is important to distinguish between a code of ethics that guides planning practice and the moral foundations that support plans and policy-making. In our discussion, attention is directed at the ethical foundations that guide our thought and decisions. The reason for this focus is based on the view that the social allocation of land to different uses and activities is fundamentally a problem not about codes of conduct, but of moral judgment (Beatley, 1994). Planning decisions have, individually and cumulatively, inescapable social and environmental impacts that range from the destruction of critical

Table 9.1 Entities involved in deciding planning issues.

Landowners and landholders
Builders and land developers
Users of public lands
Community interests groups
Elected and appointed government officials
Planning and environmental professionals
Banks and lending institutions
Homeowners
Environmental and conservation groups

Based on Beatley, 1994.

elements of the natural systems to the initiation of development trajectories that affect factors such as resource use, socialization, and the health and safety of population groups. When looked at in this manner, how we decide affects the condition and quality of the natural and built environment and ultimately the quality of people's lives. Therefore, planning decisions are not trivial. They involve making ethical judgments and choices that raise fundamental questions concerning right and wrong, good and bad. Our general treatment of ethics and their role in policy-making now narrows down to the simple fact that ethical judgments are not optional. Rather, they are inherent to every task undertaken by the planner, and while individuals make these moral judgments, their ramifications can cut across a spectrum of social, governmental, and institutional entities (Table 9.1). Consequently, the scale and importance of the decisions we make will vary depending on factors such as: personal enlightenment, social responsibility, and the professional role we enjoy.

Theories and questions

The subject of ethical planning remains a widely debated topic. While planners and environmental professionals have long recognized the ethical nature and importance of policy decisions, efforts at espousing a set of planning ethics have not been forthcoming. Much of the problem has been the lack of connection between the ethics that have been offered and specific moral theories or concepts that can help to defend of justify them. An ethic may incorporate exhortations and encouragements, but without a strong theoretical underpinning such statements will be questioned (Beatley, 1994). Furthermore, environmental ethics are often presented at such a general level that their practical meaning can be extremely vague. As Beatley (1994) observes, what exactly does it mean when a policy dictates that land-use decisions must protect the interests of the least advantaged members of society, or when a decision must protect the integrity of a natural system? Concepts such as least-advantaged and integrity are subject to interpretations whose meanings are inexact.

As with traditional moral philosophy, planning ethics are also concerned with concepts such as obligations, rights, social justice, and virtue with a focus aimed squarely at what ought to be rather than what is. However, planning ethics also assumes that individuals have the capacity to reason, contemplate, engage in moral judgment, and modify our decisions and behaviors according to the outcomes of such judgments. Arriving at an ethic demands consideration of a number of critical moral and ethical questions unique to the environmental planning problem (Beatley, 1994):

Defining the relevant moral community – The moral community defines those individuals that stand to have their well-being or interests affected by our policy decision. There are three important dimensions that influence how the relevant moral community is determined. The first is a geographic dimension that requires us to consider the spatial characteristics of the community and its relative scale. For a given policy decision, is the planner obligated to consider its ramifications with respect to the immediate neighborhood, or should broader interests be included that extend to the regional, state, or national scales? The temporal dimension follows next. Given the realization that policy decisions can have effects that carry into the future, the temporal dimension asks us to consider how far into the future our obligations should extend. Are we obligated to consider only the present population or must we think in cross-generational terms? Since a policy decision may introduce cross-generational impacts, are we obligated to include a much longer time horizon and factor into our moral calculations the effects of an action well beyond the present? If there are in-

tertemporal obligations, how far into the future must we consider? The final dimension to consider when defining the relevant moral community is biological. When making policy decisions, to what extent must we consider the interests and well-being of other forms of life?

Defining ethical obligations – While reaching a definitive conclusion regarding the extent and parameters of the moral community may prove challenging, once a judgment has been made the actual ethical obligations owed to the various members of this community can be examined. Here, ethical principles, standards, and concepts from moral philosophy and environmental ethics can be consulted. Thus, given a policy question, is it our moral obligation to make a decision based on utilitarian and economic concepts, or do we have obligations to prevent actions that impose harm and promote actions that protect the rights of individuals? Do we have an obligation to keep our promises (Beatley, 1994)?

Defining the moral grounds of our ethical standards – Supporting our standards involves the reasoning and methodologies we can employ to arrive at and defend an ethical principle. Defending and justifying ethical concepts is controversial and often contentious; however, those involved in policy-making will be called upon to defend their positions.

Making decisions – Decision-making is about choices and in the process of choosing alternative ethical standards, ethical concepts and their implications must be weighed and evaluated. The questions that arise here typically concern differing ideas of what constitutes participation in decision-making, the role of individuals affected by the process, and issues related to their representation in this process.

These rudimentary questions remind us that environmental planning and the ethical circumstances surrounding the subject are too varied and complex to invite a single, unified approach. Instead, different situations calls into play different subsets of ethical concepts and principles (Beatley, 1994). Categorizing the different ethical principles central to environmental planning can be accomplished by considering the extent to which they fall along the teleological/deontologi-

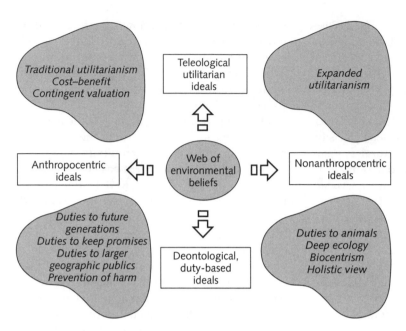

Fig. 9.1 The web of environmental beliefs.

cal–anthropocentric/nonanthropocentric continuum (Fig. 9.1).

A teleological theory states that the basic or ultimate criterion of what is morally right, wrong, or obligatory, is the nonmoral value that is brought into being (Frankena, 1973). With such a theory, concern is given to the comparative amount of good produced or the comparative balance of good over bad. When translated into the language of planning, teleological principles direct judgment regarding matters of policy to encourage actions that generate the greatest degree of good or value over bad. Therefore, the appropriate decision is the one that will maximize what is good. Of course, the issue to resolve is the problem of whose good is to be maximized. If an ethical egoism is applied, then the morally correct action is the one that maximizes value for a specific individual. Alternatively, one could select a utilitarian approach that applied maximization to the collective good. Presently, contemporary policy-making in planning is based on utilitarian principles that can assume several different forms:

- traditional utilitarianism
- cost/benefit analysis
- market failure
- contingent valuation.

At the opposite end of the continuum are those ethical principles that are described as deontological. Deontological principles reject the assumption that the morally correct action is necessarily the one that maximizes good. Rather, deontological ethics assert that there are other considerations that may make an action right or obligatory besides goodness. These other considerations include the requirement that the action or policy keep a promise, be just, or be commanded by an authority. Some examples of deontological ideals in planning include:

- land-use rights
- culpability and prevention of harm
- distributive ethics and social justice
- duties to future generations
- duties to a larger geographic public
- duties to keep promises.

Disagreements over planning policies often result from a conflict between teleological and deontological points of view (Beatley, 1994). We can examine the question of re-zoning land to illustrate how conflict may emerge. Suppose a request was made to re-zone a parcel of land in a residential area to permit the operation of a convenience store. The teleological view would require evaluation of the ethical merits of the proposed zoning change by examining the net benefits such a change would bring to the community. If the change can be shown to generate more benefits than costs, the change in zoning would be justified. Using this same example, suppose a neighboring homeowner, who would bear a disproportionately large degree of negative consequences from excessive noise and traffic that would be associated with the new land use, recently bought his home with the understanding from the city that no such uses would ever be permitted in area. The homeowner now feels betrayed. Deontological principle would move decision-makers to consider their promise to the homeowner, and view their moral obligation to keep a promise as more important than the benefits that would accrue to the neighborhood from the zoning change.

Another critical dimension into which many ethical positions fall concerns the extent to which moral obligations are human-centered (anthropocentric) or nonanthropocentric. For example, conventional utilitarianism generally assumes a human-centered perspective since the costs and benefits associated with the relevant decision criteria are those that affect or accrue to human beings. There are, however, various nonanthropocentric obligations that have been expressed within the fabric of environmental ethics, including:

- duties to animals and sentient life
- duties to protect species and biodiversity
- biocentrism
- deep ecology.

The question as to whether moral obligations are human-centered or should be extended to include nonanthropocentric considerations carries important environmental policy implications. To illustrate this point, examine a sample of questions that might be raised as a policy-maker reviews plans to accommodate regional growth:

- Do we owe obligations to other forms of life and if so, which ones?
- Do we owe obligations only to sentient lifeforms, or only to lifeforms with relatively high levels of intelligence?
- Do we owe obligations to broader categories of life?

- Do we owe obligations to the broader ecosystem or ecological community irrespective of the value it might hold for human beings?

Answers to these questions are not obvious and typically can be restated in terms of what we identify as "value." Three qualities embedded in environmental policy can be looked at as possible ways to express our preferences: (1) instrumental value, (2) intrinsic value, and (3) inherent value. Instrumental and intrinsic values share an anthropocentric nature since they rely on people to express their attitudes regarding conditions of the environment. For example, a forested area can hold both an instrumental and intrinsic value to humans. The area's instrumental value may be found in the forest's ability to supply trees that satisfy a human need. Its intrinsic value develops from its recreational and aesthic qualities and suggests that the landscape provides us with an important connection to nature. A more complex idea is the concept of inherent value. Inherent value implies that regardless of the value ascribed to the landscape by humans, the landscape may have a value that is not derived from people, but because the forest is a living thing with a "good" entirely of its own. Questions of value, the connection between ethics and planning, and what concepts exist that may be called upon by the environmental planner, invite an exploration of ideals and philosophies that drive environmental thought.

Contemporary environmental thought

The questions raised by the introduction of ethical concerns in environmental planning direct us to adopt an expanded view of the moral community and rethink existing environmental attitudes and behaviors. Replacing existing attitudes with more contemporary environmental views begins by exploring how these views impact our conceptualization of human/environment relationships and how these streams of thought relate to planning theory and planning practice. As Jacobs (1995) observes, contemporary environmental philosophy partially affirms the position taken by progressive planners that to produce successful plans one needs to articulate and engage a set of deeper questions concerning the root cause of the problems we encounter. In this section a selection of contemporary ideas are examined that might move environmental plans and policies closer to the goal of becoming more effective, equitable, and sustainable.

Deep ecology

The term "deep ecology" was introduced by the Norwegian philosopher Arne Naess (1973). Naess used the term to contrast with another idea he called "shallow ecology." According to Jacobs (1995), shallow ecology is what would be popularly thought of as the legislative management orientation of mainstream environmental planning and management. Naess maintains that this "shallow" orientation is flawed and asserts that as long as management focuses upon reforms and "at-the-margin" tinkering with an industrial-technological society where people relate to nature in a utilitarian and anthropocentric fashion, society will never realize a truly sustainable way of living with the earth. Deep ecology is a profoundly biocentric worldview that calls for a fundamental shift in our attitudes and behaviors. Biocentric worldviews that have emerged from deep ecology are now crystallized in a set of ideals that have become known as the Earth-wisdom worldview.

The Earth-wisdom perspective is based on the following major principles that stand in sharp contrast to the more traditional views of environmental management:

1 Nature exists for all of the Earth's species not just for humans, and humans are not in charge of the rest of nature.
2 There is not always more, and what there is, is not all for humans.
3 Some forms of economic growth are environmentally beneficial and should be encouraged, while some are environmentally harmful and should be discouraged.
4 Our success depends on learning to cooperate with one another and with the rest of

nature, rather than attempting to dominate and manage the Earth's life-support systems primarily for human use.

Therefore, according to the tenets of deep ecology, to achieve an environmentally sustainable world, people have to acknowledge and afford equality to all living creatures. Thus, the richest and most just form of life on Earth is a broad state of species and social organizations, diversity and complexity (Jacobs, 1995). This dual focus on diverse and complex social organization and more nature-based human societies encourages deep ecologists to advocate local autonomy and decentralized forms of social and political organization. Recognizing that all life has value on its own terms, deep ecology sees a natural wisdom to the organization and functioning of ecological systems that have not been disrupted by human activities. This natural wisdom can be learned by observing wild nature, preserving and protecting natural areas, and affording credibility to the knowledge of more nature-based human societies.

Ecofeminism

The term ecofeminism, coined in 1974 by the French writer Françoise d'Eaubonne, includes a wide spectrum of views on the relationships of women to the Earth and to male-dominated societies. In a manner similar in concept to deep ecology, ecofeminism begins with a concern with the ability of the shallow environmental movement to solve environmental problems and asks for a deeper analysis and understanding of Earth-centered environmental worldviews. The fundamental theme of ecofeminism is the belief that a central cause of our environmental problems is not simply human-centeredness, but more specifically male centeredness. Thus, the roots of oppression of animals, plants, and landscapes and the roots of oppression of women are inextricably linked. According to ecofeminist theory, the rise of male-dominated societies and environmental worldviews since the dawn of agriculture is primarily responsible for our violence against nature and for the oppression of women and minorities. Patriarchal cultures validate male-associated values and denigrate that which is nonmale. Such values brought into question by ecofeminist theory include an excessive reliance on rationalism, dualistic forms of intellectual organization, hierarchy as a mode of conceptualization and organization, and rule-based models of management (Jacobs, 1995).

Based on an ecofeminist perspective, the roots of liberation of women and nature are connected by a common need to reform how we understand, think, and conceptualize the world. Therefore, an internal, conceptual reorganization and reorientation is needed that validates alternative explanations of knowledge and management. According to Jacobs (1995) these include concepts that stress intuition, interconnected and systems forms of organization, nonhierarchical forms of organization, and process and equity forms of management. When placed into the realm of environmental planning, ecofeminist ideals challenge the process of how things are done, from the equity issues surrounding administrative hearings to the manner by which environmental impact statements are produced.

Bioregionalism

Bioregionalism is a concept that originates from a careful observation of the earth, its patterns, and the manner by which people accommodate and become part of those patterns. When compared to deep ecology or ecofeminism, bioregionalism does not arise from questions surrounding the relationship between people and the natural world, and concerns regarding flawed worldviews. Rather, bioregionalism is concerned with how people live in a place, and how people learn from living in that place (Andruss et al., 1990). Working with and planning for the Earth, according to bioregional thinking, is to view the geographic area where we live as part of a unique life territory with its own soils, landforms, watersheds, microclimates, native plants, native animals, and other natural features that make the region distinctive. This view defines the concept of a bioregion that has come to be realized as a geographic terrain and a terrain of consciousness. Within a bioregion, the conditions that influence life are similar and

this in turn influences the form and function of human occupancy (Berg & Dasmann, 1978). The central qualities of a bioregion are: (1) a distinct ecospace, distinguishable from other ecospaces, (2) a distinct form or style of human use that reflects the influence and power of the land on settlement.

The obvious question when exploring the topic of bioregionalism is how a system-ecological definition can give rise to an environmental philosophy. Here, the significance of bioregionalism is not simply that of thinking about the environment or acting to prevent environmentally destructive behavior, but in learning to live in such a way as to gain knowledge and an appreciation of place that profoundly affects how we live there. The underlying assumption of the bioregional perspective is that by being fully alive in and with a place, people will cease to cause environmental damage (Berg & Dasmann, 1978). Thus, by developing sensitivities to the ecosocial carrying capacity of their region, a population will learn how to use land more fully without abusing it. From a planning perspective, perhaps the most important implication of bioregionalism is that people should not live the same everywhere. Instead, development (settlement) should accommodate bioregional variations and urban and rural form should differ in response to bioregional influences.

Most attempts at bioregional living involve reinhabitation: learning to live in an area that has been disrupted by human exploitation. In these instances, reinhabitation focuses on:

- What the region was like before human settlement introduced change.
- Identifying what natural trajectories would have produced natural form in the absence of human intervention.
- Determining strategies to rehabilitate the bioregion, cooperating with and re-establishing the natural processes that shape and sustain it.

Holistic ethics and land duties

Developing from an expanded interpretation of Leopold's land ethic, holistic ethics maintains that obligations are owed to ecosystems solely because of their complexity, uniqueness, and their intrinsic value. As Rolston (1988) suggests, when humans awaken to their presence in the biosphere, finding themselves to be products of the processes that define it, they owe something to maintaining its integrity and persistence. Of course arguments to protect natural environments are compelling; defending these arguments apart from human self-interest is more challenging. Moving beyond the benefits to humans, a set of land duties can be offered that direct ethical environmental planning regardless of biological level. These ethics introduce requirements that:

- Efforts be made to minimize the extent of damage and destruction to the natural environment.
- Acceptance of impacts be permissible only for important social purposes.
- Efforts be made to promote the recovery and re-establishment of populations, habitats, and damaged ecosystems.
- Serious moral consideration of the interest of other forms of life and the habitats on which they rely be encouraged.
- Consideration be given to minimizing the extent of the human footprint on the natural system.

Environmental justice and equity

In the preceding sections we've examined the concepts of rights and obligations within the context of the broader environment. Contemporary thinking seems focused on reformulating egocentric paradigms of human/environmental relationships and introducing higher levels of environmental awareness, encouraging more biocentric/ecocentric worldviews. While there is great value in developing environmental policies that shed human-centeredness, there is the underlying assumption that all humans share the same level of environmental quality and enjoy concomitant rights to a healthful and productive biosphere. Based on our previous discussion, everyone has a fundamental right to clean air and clean water, and no one has the right to degrade or destroy the places in which we live. However, within the last decade a growing body of evidence suggests that many environmental policies,

directives, and actions differentially affect or disadvantaged individuals, groups, and communities based on race, ethnicity, or socioeconomic status. For example, in a landmark study undertaken by the Commission for Racial Justice in 1987, it was shown that:

- The greatest number of commercial hazardous facilities were located in communities with the highest composition of racial or ethnic minorities.
- The average minority population in communities with one commercial hazardous waste facility was twice the average minority percentage in communities without such facilities.
- Race was the most significant variable associated with the location of hazardous waste sites.

In a study conducted by the US Environmental Protection Agency (1992), findings showed that socioeconomic conditions and race are the major factors that influence environmental discrimination and that communities inhabited by poor whites are also vulnerable. These observations have propelled the environmental justice movement into the forefront of environmental policy-making, and have introduced a new set of considerations that must be understood by the environmental planner.

In general, environmental justice has been defined as the pursuit of equal justice and equal protection under the law for all environmental statutes and regulations without discrimination based on race, ethnicity, or socioeconomic status. As a compelling interest, environmental justice seeks to dismantle exclusionary zoning ordinances, discriminatory land-use practices, differential enforcement of environmental regulations, and the disparate siting of risky technologies. In their place, environmental justice endeavors to encourage greater equity in policy-making (Bullard, 1995).

As a force for change, environmental justice embodies five central and interrelated themes:

1 Environmental Equity – an ideal of equal treatment and protection for various racial, ethnic, and income groups under environment statutes, regulations, and practices ap-

plied in a manner that yields no substantial differential impacts relative to the dominant group. Although environmental equity implies elements of "fairness" and "rights," it does not necessarily address past inequities or view the environment broadly.

2 Environmental Justice – the fundamental right to a safe, healthy, productive, and sustainable environment for all, where the environment is considered in its totality to include the ecological, physical, social, political, aesthetic, and economic environments, Environmental justice refers to the conditions in which such a right can be freely exercised, whereby individual and group identities, needs, and dignities are preserved, fulfilled, and respected in a way that provides for self-actualization and personal and community empowerment.

3 Environmental Racism – racial discrimination in environmental policy-making, enforcement of regulations, and laws, and targeting communities of color for toxic waste disposal and hazardous industry siting.

4 Environmental Classism – the results of and the process by which implementation of environmental policy creates intended or unintended consequences that have disproportionate impacts (adverse or beneficial) on lower-income persons, populations, or communities. These disparate effects occur through various decision-making processes.

5 Environmental Justice Community of Concern – a neighborhood or community composed predominantly of persons of color or a substantial proportion of persons below the poverty line that is subjected to a disproportionate burden of environmental hazards or experiences a significantly reduced quality of life relative to surrounding or comparative communities.

The emergence of environmental justice concerns stems from the perception that environmental policy-makers have ignored the effects of pollution and environmental hazards on people of color, the poor, and the working class. This is seen largely as a consequence of the lack of politi-cal power in those communities and their inability to mount a serious protest against such hazardous activities. Studies conducted to examine racial and demographic variables and environmental quality supply compelling evidence of this claim, and the ethical questions they raise carry important implications for how policy and plans are made. Given the observed pattern that poor people lack the economic means to leave their neighborhoods for resettlement elsewhere, and that housing discrimination often makes it difficult to find alternative, affordable housing, coupled with the realities that hazardous industries are attracted or relegated to areas of low land value where the poor tend to reside, a cycle of inequity is established that become easy to replicate elsewhere. This pattern is further strengthened by the structural barriers that characterize poor areas and their lack or community resources to prevent or control the entry of risky industries into their community (Mohai & Bryant, 1992).

Dealing more equitably with minority groups and the poor in environmental planning requires sustained efforts to:

- Increase the priority given to issues of environmental equity.
- Improve the risk assessment procedures to ensure better characterization of risk across population, communities, and geographical areas.
- Increase efforts to communicate with minority and low-income communities and involve these groups in the policy-making process.
- Establish mechanisms to include equity considerations in long-term planning.

Realizing these objectives places emphasis on purposefully connecting ethical principles regarding the total environment with the planning process.

Linking thought to planning practice

The environmental philosophies examined in this chapter carry important implications for planning and the general process followed in the formulation and implementation of environmental plans.

Perhaps the most significant realization suggested by these ideals is their insistence on the need to broaden the scope of planning and extend considerations beyond the typical motivations that have traditionally guided the planning process. In practical terms, the message is simple: it is not useful to make little plans or to view planning as an incremental process. Piecemeal approaches may be counterproductive and delay examination of the underlying and fundamental causes of problems (Jacobs, 1995). Only by broader, deeper examination will long-lasting and sustainable solutions emerge.

The philosophical and ethical issues reviewed in this chapter can also lend validity to both rational and progressive planning by raising important questions that expose the structural origins of the conditions that underlay the world, how we view the world, and how we respond to its challenges. Such questioning should facilitate deliberation and understanding in a long-term, systemic context. By thinking about the questions posed and responding to them and the conditions they define, more comprehensive strategies can emerge that can encompass a wider description of the problem. Simply by becoming aware of these philosophical perspectives and how they provide for alternative worldviews, the planner will become more accustomed to thinking with these concepts and incorporating some of their ideals toward real solutions.

Complementing these direct influences are a series of indirect effects that stand to challenge existing planning theory. These include challenges relating to (after Jacobs, 1995):
- The legitimacy of an abstract or contextless planning theory.
- The general anthropocentric orientation of planning theory and practice.
- The relationship of means and ends.
- The loss of "place" as a specific basis for planning.

With minor exception, planning theory is primarily concerned with how to plan. This focus need not be connected to the actual practice of planning; therefore, much of planning theory is abstract and generalized. The ethical questions born out of the environment, while expressed in theoretical terms, originate from actual environmental problems. Thus, the contrast between planning theory and environmental philosophy is simply the difference between the question of how to plan, versus knowing what to plan for (Jacobs, 1995). For example, looking at planning through the lens of deep ecology it would seem entirely utilitarian and anthropocentric in its orientation. To those influenced by deep ecological thinking, planning would need to stress the rights of nonhumans and restructure the planning process to include nonhuman considerations. From an ecofeminist perspective, the relationship between means versus ends would be called into question; incorporating this view would challenge the entire planning process, how communication in that process evolves, the forums provided to facilitate communication, and the individuals who are empowered to act as representatives in this process. If an ecofeminist view is adopted, then planning would follow a more democratic conception of knowledge and expertise and the planner would be asked to acknowledge a broader basis for understanding the nature of a given problem.

Bioregionalism, with its focus on learning to live in and to know "place," and to be affected by its uniqueness, would direct planning to consider the evolution of life patterns and their fit within the ecological conditions of the landscape. By stressing the importance of local knowledge of place and its use in all aspects of design, planning would become grounded in "place" and less abstract. Bioregional thinking would stress the incorporation a specific spatial element (the region) into all planning analysis and use the region as a framework to support planning theory.

Equity and the distribution of risk and benefit are the primary concerns brought to planning theory from the environmental justice arena. From this perspective, planning speaks to the values exercised by the planner and the biases inherent to the planning process that discount equal protection and equal access. By giving greater emphasis to cultural diversity, social and environmental issues that differentially separate people across ethnic, racial, and economic lines, environmental justice considerations would realign planning, placing equity at the center and giving greater priority to the mix of social and environmental issues and the patterns of risk and benefit they represent. Through this heightened sense of balance access to the policy-making process can be improved, particularly for those communities with limited political and economic resources.

The direct and indirect influences contemporary environmental thinking has on the practice of environmental planning point to a number of challenges that will continue to shape and reshape how we plan, why we plan, and who we plan for. If these ideals do nothing else, they force us to look at the world differently, ask questions, and seek alternatives. For this reason alone, contemporary environmental thought may offer opportunities to develop a new body of environmental planning theory, one that is more inclusive and responsive to the environment in total. Above all, however, the ethical ideas introduced here will influence how we look at and make decisions, and how well our decisions move us toward an effective, log-term, sustainable, and equitable solution to the problems surrounding human/environmental needs and responsibilities.

Summary

As a type of decision-making, environmental planning requires a sensitivity to the issues that underlie any given planning problem and the de-cisions that must be made that concern them. In this chapter the concept of ethics and the role of ethics in environmental planning were discussed. The ethical principles planners may call upon to provide insight and guidance centered around fundamental questions of right and wrong and the ability to understand good versus bad decisions. The treatment of root ethical principles led to a discussion of contemporary environmental thought and introduced new directions that are challenging traditional point of views. Here, ideas concerning deep ecology, bioregionalism, and environmental equity were examined and their influence in reshaping human/environmental relationships was explored. Using these ideals as a backdrop, the connection between planning practice and ethics was evaluated.

Focusing questions

What are ethics and how do ethical principles help us decide?

What is the value of ethics to the environmental planner?

Identify the ethical challenges that confront contemporary environmental planning and suggest how they may be approached.

Who are the relevant moral communities in your region and what characteristics identify their worldview?

CHAPTER 10

The Impact of Change

A central assertion throughout this text has been that environmental planning is a proactive and future-oriented discipline where environmental information is blended with traditional planning concerns in order to arrive at a sustainable balance between human need and environmental quality. In this text we have also set out to demonstrate that environmental planning is a form or decision-making where a choice must be made among alternatives in relation to a set of goals and objectives that detail a desired future. Because of our emphasis on the future, given the recognized fact that this future is largely unknowable, embarking on the appropriate course of action is shrouded in an envelope of uncertainty. Although uncertainty cannot be eliminated, it can be understood and reduced. In this concluding chapter we will explore a family of procedures designed to assess change in the land-use/environment system, and explore avenues whereby environmental concerns can be better incorporated into the public-policy making arena. Our exploration begins with the environmental impact assessment process.

Environmental impact assessment

Environmental impact assessment has been characterized as the process of identifying and evaluating the consequences of human actions on the environment, and developing procedures for mitigating those consequences that are adverse (Erickson, 1994; Marriot, 1997; Canter, 1996). From its inception, environmental impact assessment (EIA) has been guided by one overriding goal: to incorporate into the decision-making process the consequences of human activities that were not being addressed adequately by the free-market exchange system (Lein, 1989). While the details of the EIA process, the technical guidelines, administrative procedures, and legal responsibilities that pertain to it vary from nation to nation, and within nations, a consensus is building regarding the fundamental elements that constitute EIA (Erickson, 1994). First in importance is the fact that EIA requires seeing the environment as the aggregate of things and conditions that surround every living and nonliving thing. Within the context of EIA the environment is not simply those nonhuman things and processes, but also includes the human world and the elements, processes, and conditions that pertain to people. Secondly, EIA is a decision-making tool whereby the possible and probable consequences of a human action are carefully examined before an irreversible commitment is made that will contribute to an adverse environmental change. Thirdly, EIA directs analysis on both the quantifiable and qualitative aspects of the environment. Here inclusion of the non-measurable qualities is critical. Conducting an assessment of impact to consider only those characteristics of the environment that are quantifiable omits critical factors that may influence decision-making and presents an incomplete view of the environment. Finally, EIA places emphasis on the concept of mitigation and asks for the detailed investigation of methods and alternatives that will reduce undesirable effects and enhance those effects that are beneficial.

Historically, when projects or policies were designed, little direct consideration was given to their possible environmental effects. Generally, environmental considerations entered into the decision-making process well after a project was underway, if at all. Decision-makers typically assumed a reactive posture when the environment was factored into their plans. However, as the cost of environmental damage became more evident it became incumbent upon society to examine its actions and attune them so they would assure the long-term viability of Earth as a human habitat. This responsibility has been embodied in, for example, the US National Environmental Policy Act and its numerous state and international successors. When the National Environmental Policy Act (NEPA) was signed into law on January 1, 1970, a significant step was taken in the US to change government decision-making and place the environment directly into the decision-making process. Following the enactment of NEPA, the concept of environmental impact was no longer a subject for academic discourse, but a physical reality that became government's responsibility to identify, manage, and control.

By definition, an environmental impact explains any alteration of environmental conditions or the creation of a new set of environmental conditions adverse or beneficial, caused or induced by an "action" or set of actions. Two terms in this definition are worth closer examination. First, consider the term "alteration". If we go to the dictionary, the term implies the act of making something different. It does not suggest how, where, or why, nor does it place a value judgment on whether the difference is good or bad. The term is purposely nonspecific and suggests that decision-makers look seriously at the nature of their proposals and explore conditions in their broadest context. Similarly, the use of the term "action" is also necessarily vague. An action can include a range of govern-

mental activities such as the allocation of money, the installation of a program, the implementation of a policy, the granting of a permit, the creation of a new physical structure or plan, or the modification of an existing plan or design.

The impacts produced by these actions can be either primary or secondary. This distinction is important because it recognizes the interconnectedness of the environment and the flow of process (impact) through a system (environment). Primary impacts define alterations of environmental conditions that can be attributed directly to the proposed action. This category of impact might be considered the "first-order" changes that result as a consequence of the action and identifies those effects that are immediate and obvious. In contrast, secondary impacts explain indirect or induced changes that result as a consequence of the action or its primary effects. Alterations of this type might be viewed as the "second-order" changes resulting from the project's perturbations.

A major aim of EIA involves the inclusion of environmental amenities into government decision-making. To incorporate amenity considerations into this process it becomes necessary to develop a complete understanding of the possible and probable consequences of the proposed action. Accomplishing this task suggests that EIA is more than a set of ideas or a terminology, but a method that seeks to:

- Develop a complete understanding of the proposed action.
- Gain a complete understanding of the affected environment.
- Project the proposed action into the future to determine its consequences.
- Report on the nature of the projected consequences attributable to the action in a manner that facilitate an informed trade-off between the action and its alternatives.

Fundamental to this method of analysis is the ability of the analyst to read the landscape and mentally overlay the action on its environmental setting. In the context of EIA, the environment is viewed as the whole complex of physical, social, cultural, economic, and aesthetic features, and for each the expected impacts are identified in relation to the project and its features.

Today, EIA is the embodiment of a systematic methodology for proactive decision-making that sets out to document, predict, and monitor human actions. Although EIA has a history dating back three decades, it remains very much an evolving methodology comprised of a mix of analytic techniques and administrative procedures that contrast in terms of purpose and comprehensiveness (Marriot, 1997). Because EIA continues to evolve there is considerable interest in improving assessment, with efforts concentrating on five main themes (Bartlett & Malone, 1993; Canter, 1993; Malik & Bartlett, 1993):

1 The development of analytic models used to guide the prediction of environmental consequences.
2 The refinement of procedures used to identify and reduce the effects of uncertainty.
3 The incorporation of new techniques that can accommodate subjective judgments and qualitative data.
4 The inclusion of cumulative impacts into the assessment.
5 The design of environmental monitoring programs for long-term project management.

Taken together, these themes identify common characteristics that should encourage more integrative approaches to EIA (Lein, 1998).

Perhaps the most important concept guiding EIA is that of disclosure. As a decision-making tool, EIA is, in essence, a process of bringing the environmental consequences of a proposed action to the attention of both decision-makers and concerned members of the public, making those effects known and providing the opportunity for comment and review of their significance. Theoretically, disclosure unfolds as a continuous sequence of analytical stages that culminate with the formal documentation of effects. Documentation takes the form of a written report that serves as the primary decision tool and communication device that connects the proposed action to all concerned parties. While the assignment of EIA to a single sequence of analytical procedures may not be a suitable generalization in all circumstances, its conceptual foundation is firmly rooted in the EIA planning process (Lawrence, 1994). Barrett and

Therivel (1991) have defined this ideal as a unifying approach to impact analysis that:

- Applied to all projects that are expected to have a significant environmental impact.
- Compares alternatives to proposed project management techniques and mitigation measures.
- Results in a clear documentation of effects that conveys the importance of the likely impacts and their characteristics to experts and non-experts alike.
- Includes broad public participation and stringent administrative review procedures.
- Is timed in a manner that provides information for decision-makers and is enforceable.
- Includes mechanisms for post-EIA monitoring and feedback.

Realizing this ideal, however, underscores the contrast between EIA theory and practice (Lawrence, 1994). The practice of EIA has been substantively aided by a heightened emphasis on the use of scientific principles, procedures, and knowledge that recognizes the importance of explicit study design, assumption testing, and the clear establishment of spatiotemporal boundaries (Beanlands, 1989; Malik & Bartlett, 1993). The focus on EIA as science has contributed to the view of impact prediction and measurement as a set of rigorous and reproducable techniques guided by the studied application of expert judgment and experience (Lawrence, 1994; Lein, 1993b). Viewing EIA as science reduces the problem of assessment to four general considerations:

1 Careful selection and use of multiple social and natural environmental indicators.
2 Detailed concern for mathematical validity.
3 Applications of statistical tests of significance.
4 Sensitivity to data reliability, systemic bias, and nonlinear relationships.

It has been argued, however, that viewing EIA as a pure prediction of events is an unwise and unrealistic standard, since such a mindset tends to ignore the stochastic nature of human and environmental processes (DeJongh, 1988). Adhering to the principle that EIA is solely prediction tends to minimize the presence of uncertainty and encourages the assumption that a model exists to characterize every aspect of the assessment problem and that a single planning process can be applied uniformly to all projects. Lein (1993) discussed the implications of these assumptions, noting that prediction when applied to EIA implies the ability to foretell with bounded precision the future outcome of a given course of action. Reality suggests, however, that subjective-technical judgment and qualitative approaches are common to the practice of EIA. Therefore, conducting EIA in a purely predictive mode obscures the role played by expert judgment and discounts the inclusion of effects that cannot conform to this type of normative problem-solving. To some practitioners, expert judgment may seem to contradict the concept of EIA as science, yet its presence alludes to the realization that quantitative prediction may, in certain instances, be impossible, impractical, or inefficient (Lein, 1993b). Therefore, while a perfect quantification may not be a necessary condition of EIA, a structured problem-solving approach remains essential.

The method of EIA

Environmental impact assessment involves five basic activities:

- **Impact identification** – defining the task of identifying the impacts that need to be investigated. Although this seems like a relatively straightforward problem, there is a lack of knowledge concerning the nature and extent of impacts that frustrates this seemingly simple task. The environmental consequences that may be associated with an action in one location may be very different from an identical action located in another environmental setting. Thus, impact identification can be complex and continuous, yet before proceeding with a detailed assessment, preliminary identification of impacts is an absolute necessity. This early identification places a premium on project scoping and screening activities as a means of collecting information on the nature of the project and its environmental setting, and narrowing down the scale of an assessment to select those environmental factors critical to an understanding of the project's consequences.

- **Impact measurement and prediction** – this phase of EIA involves estimating the potential nature of environmental impact associated with the action expressed in quantitative and/or qualitative terms. Frequently, the magnitude of the possible changes attributed to a project must be predicted quantitatively. These predictions can be obtained in several ways, including the use of mathematical models, physical models, or computer simulations. Measuring the changes in the state of environmental factors is the first step in estimating the nature of impact. Once these measures are obtained they can be associated back to humans, plants, or animals to specify the exact nature of their effects in relation to these receptors.

- **Impact interpretation** – interpretation defines two distinct operations: (1) determining the significance of an impact and (2) evaluating the magnitude of an impact. For EIA to function as a decision tool, a clear distinction between magnitude and significance is necessary. In general, the magnitude of an impact is arrived at by prediction based on empirical measurements, significance tends to develop as an expression of the "cost" of the predicted impact to society (Thompson, 1990).

- **Impact communication** – following interpretation, the quantitative and qualitative information describing the impacts attributed to the proposed action need to be documented and presented in a form that enables experts and non-experts to understand and comprehend them. Because EIA carries the responsibility of full disclosure, effective communication is an essential ingredient. Unless decision-makers and concerned members of the public can understand the assessment, informed conclusions cannot be reached and the relative merits and risks associated with the project cannot be fairly evaluated.

- **Impact monitoring and mitigation** – minimal attention has been given to comprehen-

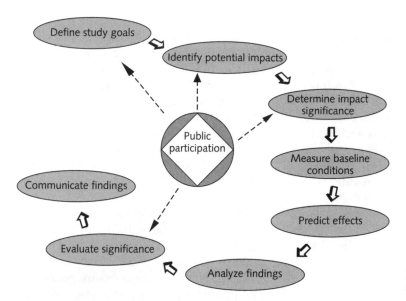

Fig. 10.1 The general method of environmental impact assessment.

sive environmental monitoring in conjunction with the EIA process (Canter, 1993). Three forms of environmental monitoring can be employed throughout the life-cycle of a project: (1) baseline monitoring, which defines the measurement of environmental variables to help determine existing conditions, (2) effects monitoring, which describes the measurement of environmental variables during project construction and operation to signal changes caused by the action, (3) compliance monitoring, which explains the periodic sampling or continuous measurement of project behavior to ensure developmental conditions are observed and environmental performance standards are met.

The five elements of EIA outlined above fit into a highly stylized procedure that suggests the sequence of tasks that are performed when conducting an EIA (Fig. 10.1). Beginning with the definition of goals, project assessment moves in an interactive fashion through the identification, measurement, and prediction phases, and culminates with the evaluation of significance and the documentation of effects. Within each of these phases, a series of subtasks can be described that complete the detailed methodology that has become the EIA process (Table 10.1). Typically, it is

Table 10.1 The phases of EIA.

Phase 1	*Organization/Screening and Scoping*
	• define study goals
	• identify potential impacts
	• determine impact significance
Phase 2	*Measurement and Analysis*
	• measure baseline conditions
	• predict effects of actions
	• estimate likelihood of predictions/forecasts
	• summarize and analyze findings
Phase 3	*Evaluation and Review*
	• evaluate findings
	• examine alternatives
	• evaluate mitigation strategies
Phase 4	*Communication and Documentation*
	• describe findings
	• communicate results
	• decide on proposed action
Phase 5	*Post-Impact Review*
	• implement monitoring program
	• modify/mitigate actions

within these more detailed aspects of EIA that the link to decision-making can be understood. However, regardless of the type of project involved, the goals directing EIA remain constant: to identify impacts, to evaluate their importance, and to communicate concerns regarding the project and its alternative. To understand these goals we can return to the pivotal legislation that formalized EIA, the USA National Environmental Policy Act (NEPA).

EIA and NEPA

Environmental impact assessment has evolved and diffused from its initial beginnings, yet the US National Environmental Policy Act of 1969 remains as the cornerstone for the approach to the management of change (Clark, 1997). The National Environmental Policy Act of 1969 (NEPA) was the first legislation of its kind to be adopted by any national government and has been widely emulated; however, few statutes are as intrinsically important and so poorly understood as NEPA (Caldwell, 1997). The fundamental purpose of NEPA was twofold. First, NEPA declared a national policy to protect and promote environmental quality. As stated in title I, section 101 of NEPA, the US federal government is charged with the responsibility to use all practicable means and measures to foster and promote the general welfare, to create and maintain conditions under which [man] and nature can exist in productive harmony, and fulfill the social, economic, and other requirements of present and future generations. From this broad statement of policy, the goals that direct EIA were enumerated:

- To fulfill the responsibilities of each generation as trustees of the environment for succeeding generations.
- To assure for all Americans safe, healthful, productive, and aesthetically and culturally pleasing surroundings.
- To attain the widest range of beneficial uses of the environment without degradation, risk to health or safety, or other undesirable and unintended consequences.
- To preserve important historic, cultural, and natural aspects of our national heritage, and maintain wherever possible an environment which supports diversity and variety of individual choice.
- To achieve a balance between population and resource use which will permit high standards of living and a wide sharing of life's amenities.

- To enhance the quality of renewable resources and approach the maximum attainable recycling of depletable resources.

These statements of policy are followed by a set of "action forcing" provisions beginning in section 102(2) of the act. Under this section, all federal agencies are required to make a full and adequate analysis of all environmental effects of implementing their programs or actions. This includes a directive requiring that all policies, regulations, and public law are interpreted and administered in accordance with NEPA, and that all federal agencies follow a series of steps to ensure that the goals of NEPA are met. This provision begins by asking that a "systematic and interdisciplinary approach" is used to facilitate the integrated use of the social, natural, and environmental sciences in decision-making. However, the most significant aspect of NEPA's action-forcing mandate is found in section 102(2)(C). Here, all federal agencies, where actions significantly affect the quality of the human environment, are required to prepare a detailed statement of environmental impact. The prescribed elements of this environmental impact statement (EIS) include assessment of

1 The environmental impact of the proposed action.
2 Any adverse environmental effects which cannot be avoided should the proposal be implemented.
3 Alternatives to the proposed action.
4 The relationship between local short-term uses of man's environment and the maintenance and enhancement of long-term productivity.
5 Any irreversible and irretrievable commitments of resources which would be involved in the proposed action should it be implemented.

These five directives establish the basis for the EIA process and suggest a generalized framework for conducting an assessment. The specifics of the EIS process were refined and expanded upon by the Council on Environmental Quality (CEQ) (an agency sitting off the executive branch created in title II of NEPA).

The Council on Environmental Quality has issues guidelines to clarify the procedures to follow when preparing an EIS, the contents to be included in an EIS, and provided definitions of terms used in the EIA process to reduce confusion and enhance communication. According to the CEQ guidelines, 8 major points are to be addressed in an EIS:

1 A description of the proposed action, a statement of its purpose, and a description of the project's environmental setting.
2 The relationship of the proposed action to land-use plans, policies, and controls for the affected area.
3 The probable impact of the proposed action on the environment.
4 The alternatives to the proposed action.
5 Any probable adverse environmental effects that cannot be avoided.
6 The relationship between local short-term uses of man's environment and the maintenance and enhancement of long-term productivity.
7 Any irreversible and irretrievable commitments of resources (natural, cultural, labor, materials).
8 An indication of what other interests and considerations of federal policy are thought to offset the adverse effect identified.

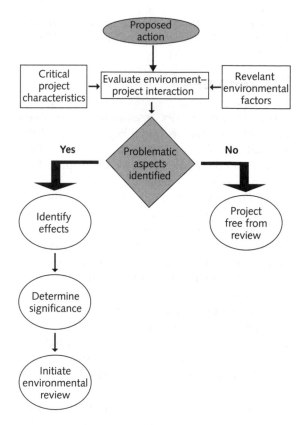

Fig. 10.2 The impact screening process.

Impact identification and screening techniques

The NEPA mandate specifically requires agencies to undertake a systematic approach to the assessment of environmental impacts and to develop methods and procedures to guide analysis. In the time that has elapsed since NEPA became law, a range of methods have been developed to assist with impact screening, identification, and forecasting (Bisset, 1988; Dickerson & Montgomery, 1993). Screening may be defined as the process of initial review that establishes a mechanism for the identification and selection of actions that require detailed analysis. As a technique, screening represents the first look at a proposed action and functions primarily to separate actions based on their probable effects. Thus, screening acts as a filter that ideally traps activities that should undergo

more careful scrutiny. The general stages followed when screening proposed actions are illustrated in Fig. 10.2. The methodologies useful for this type of identification include interaction matrices, checklists, and more recently the use of expert systems (Canter, 1996; Lein, 1989). Interaction matrices range in sophistication from simple consideration of project activities and their impact on environmental factors to more detailed approaches that display interrelationships between impacted factors. Checklists range in complexity from simple listings of environmental factors that might be relevant given the proposed action, to descriptive approaches that include consideration of impact measurement. The applicability of these methods is summarized in Table 10.2.

While the methods presented in Table 10.2 can be useful, there is no universal approach that can be applied to all project types in all environmental

Table 10.2 Overview of impact screening methods.

Activity	Method	Utility
Impact identification	Matrices: *Simple*	High
	Stepped	Medium
	Networks	High
	Checklists: *Simple*	Medium
	Descriptive	Medium
Describing affected environment	Matrices: *Simple*	Low
	Stepped	Low
	Networks	–
	Checklists: *Simple*	High
	Descriptive	
Impact prediction	Matrices: *Simple*	Medium
	Stepped	Medium
	Networks	–
	Checklists: *Descriptive*	High
	Scaling/rating	Low
Selection of alternatives	Matrices: *Simple*	Medium
	Stepped	Low
	Checklists: *Scaling/rating*	Medium
	Checklists: *Weighed*	High
Summarization	Matrices: *Simple*	High
	Stepped	Low
	Checklists: *Simple*	Medium

Based on Canter (1996).

settings. For an EIA method to be useful it should be (after Lee, 1983):

1 Appropriate to the necessary tasks.
2 Relatively free from bias and reproducible.
3 Economical and flexible.

Since EIA methods do not provide complete answers to all questions related to a project or its alternatives, selection of the appropriate method should be based on careful evaluation and professional judgment. The appropriate technique is generally that method which helps to focus thinking, provide a simple organizational structure, enable the synthesis of information, and help evaluate alternatives on a common set of criteria.

Interaction matrix methods

The interaction matrix was one of the earliest EIA methods introduced into practice (Canter, 1996). Interaction matrices are two-dimensional tables that compare environmental attributes with pro-

ject characteristics. By comparing the items listed along the row with the elements presented across the columns, project activities can be associated to the environmental factors they potentially disrupt. When a given action is expected to cause a change in a specific environmental factor, this relationship (interaction) is noted where the association intersects. The nature of the relationship can be treated by simply marking the interaction or by imposing a magnitude or importance rating. Interaction matrices fall into general classes: simple or stepped (Canter, 1996).

Simple matrix methods   One of the more widely cited examples of a simple interaction method was introduced by Leopold et al. (1971). A matrix of this design lists specfic actions against environmental items. Where an impact is anticipated, a notation is made on the matrix that identifies the association; however, in this design each action and its impact is described in terms of its relative magnitude and importance. The logic is simple. Magnitude is expressed as an intensity along a 1 to 10 scale, with 10 representing a large magnitude and 1 a small magnitude. The importance of an interaction is related to its significance using a similar scaling system. While magnitude is based on an objective evaluation of factual information, importance scores are assigned on the basis of subjective judgment.

Stepped matrix methods   Stepped matrices are commonly referred to as cross-impact matrices. These tools are often used to identify secondary and tertiary impacts associated with a proposed action. According to Canter (1996), a stepped matrix is one in which environmental factors are displayed against other environmental factors. Because stepped matrices facilitate the tracing of impacts, they are a useful means of describing causal sequences. However, as the levels of impacts and the types of actions increase, they can become difficult to interpret.

Checklist methods

Checklist approaches for impact identification range in complexity from basic lists of environmental factors to detailed descriptions that can include rankings or scaling of impact. Two checklist methods are common in EIA: simple checklists and descriptive checklists.

Simple checklist As the name implies, simple checklists are a basic accounting of the environmental factors that should be addressed in an assessment. Checklists provide a means to identify environmental factors that note areas of concern and help to prioritize and plan for more detailed studies.

Descriptive checklists Descriptive checklists provide detailed information on the measurement or prediction of impacts associated with environmental factors. For each item in the list, information is presented on possible environmental effects. In some cases, factor definitions are given and prediction methods are included.

In general, checklists offer a structured approach for identifying key impacts, together with background on key environmental factors that should be examined in the assessment. They also can be used to facilitate discussion and deliberation during the planning, implementation, and summary stages of the EIA process. However, to be effective, they must be carefully defined and documented, with weighting criteria and spatial boundaries clearly delineated.

Impact forecasting

Assessing the impact of a proposed action on the environment begins by projecting the action into the future and asking some fundamental questions about how the project will alter ambient conditions. Using the information gathered during the screening and scoping phase, a scenario is created that speaks to the basic "what ifs" a decision maker would need to understand before committing to the proposal. Not surprisingly, prediction is perhaps the most important technical activity of the EIA process and potentially the most confounding (Lein, 1993b). As noted previously, prediction when applied in EIA implies the ability to foretell with precision the future

outcome of a specific course of action. There is also an expectation that the action and its environmental setting can be resolved in a deterministic manner, placing a reliance on deterministic logic and introducing an unnecessary bias into an analysis.

A solution to the prediction bias, offered by Culhane et al. (1989), is to replace "prediction" with the term "forecast." Although this shift may appear purely semantic, use of the term forecast suggests that conjecture rather than true insight or knowledge is involved in every case. The use of the term forecast also provides a perspective that can better accommodate the use of expert judgment and subjectivity without rendering an assessment invalid. Unfortunately, the notion of a forecast gives the impression that vagueness and imprecision can be condoned in an EIS. However, as Lein (1992) shows, vagueness and imprecision can be modeled and quantified to fit existing assessment methodologies in a manner that assists decision-making.

There are three principal methods used to forecast environmental impacts: (1) judgmental approaches, (2) physical models, and (3) numerical models. Each has its advantages and disadvantages, and no single method is preferred in all instances.

Judgment methods

It has been argued that the use of expert opinion in EIA is neither visible nor formal. While subjective judgment is a recognized characteristic of EIA, the conditions that direct its use are not clear. Commonly, judgmental forecasting relies heavily of the ability of experts to express their opinions and on the analyst's ability to structure or quantify those opinions into something meaningful. One means used to control and direct the application of opinion is the Delphi method. The Delphi method was developed to increase the effectiveness of experts making forecasts as a group. Through this technique, opinions are obtained from experts through the use of survey instruments. Expert responses are tabulated and statistical measures are derived and used to summarize and support a group conclusion.

Physical modeling

Physical models are small-scale three-dimensional representations of a physical system. By subjecting the physical model to the conditions described by an impact scenario the analyst can witness the potential consequence and evaluate possible mitigation measures to reduce the potential adverse effect.

Numerical models

Numerical models are constructed from a combination of algebraic or differential equations that summarize process relationships defined by a human, physical, or environmental system. Typically, numerical models are based on scientific laws, statistical relationships, or some combination of both. Through the use of numerical models a simulation experiment can be created that can be used to explore changes in critical environmental factors as defined by the process models used to represent them. The analyst can then alter parameters and variables in the model to examine interactions and success of various mitigation strategies.

Impact forecasting is made difficult by the complexity of the environmental system involved and the possible range of direct and indirect effects a project and its alternatives may produce. At its best, forecasting permits a narrowing down of the uncertainty related to a proposed action and provides estimates, qualitative and quantitative, of the potential changes that an action may induce. At its worst, forecasting can mislead decision-makers by overstating or under-reporting environmental impacts, and cast doubt over the entire assessment process. Although EIA enjoys a 30-year history, in many respects it can be viewed as a science still in its infancy. As we learn more about environmental change and human-induced changes in particular, the scope and practice of EIA will naturally evolve. Presently, three areas of concern have emerged that are actively reshaping the science of EIA: (1) the assessment of growth-inducing impacts, (2) the analysis of cumulative environmental change, and (3) the treatment of time.

Growth-inducing impacts

The term "growth-inducing impact" communicates different concepts to different professional disciplines. To establish a common starting point for this discussion, we can explain the concept of a growth-inducing impact from an environmental planner's perspective. From this point of view, such an impact can be defined as the degree to which a project promotes, facilitates, or provides for the increased urbanization and development of the environment surrounding the project. The significance of this concept is that it recognizes the indirect and secondary effects attributable to a project and how those effects can contribute to a larger and potentially unwanted series of regional environmental changes. Although the primary focus of this concept is on urbanization, one can apply the principle of growth inducement to other environmental factors as well. Thus it is possible to articulate induced growth in terms of increased demand for resources, increased pollution levels, and increased infrastructure requirements stemming from the project's regional influence.

When one is applying the term in EIA, a distinction has to be made between projects or actions that *promote* growth, *facilitate* growth, and/or *provide* for growth. This distinction is nontrivial and underscores the significance of considering the larger implications of a project in its regional setting. These distinguishing characteristics also suggest that different actions can produce growth-inducing effects with varying implications. To illustrate this point, consider the following rules of thumb:

1 Nonretail employment centers are considered to *promote* growth.
2 Ancillary development, such as residential, retail, and service centers *facilitate* growth.
3 Basic infrastructure, such as water supply systems, waste-water treatment facilities, and major transportation facilities, *provide* for growth.

The induced growth process suggested by these heuristics can be placed into context by considering a simple word model that describes how growth develops and assumes a geographic form:

Construction of a large source of employment like an industrial or office complex generates jobs that result in the construction of dwelling units in the vicinity: the addition of dwelling units induces retail development to locate in close proximity and generates demand for a range of community, cultural, and religious facilities. All of this activity requires the construction of streets and highways that improve accessibility to the area. Improved access fosters continued urban development, and each additional source of employment spurs a new round of development which perpetuates the growth cycle.

Positioning a proposed action on the simple growth model is one of the first steps in analyzing its growth-inducing impact and for identifying the mitigation strategies that are available to reduce the secondary and tertiary effects it will induce. In many instances, however, the spatial expression of growth is not the only point of concern in EIA. For example, in air quality analysis, the concern is not "growth" *per se*, but emission growth. Therefore, the relationship between growth, be it expressed in economic, population, or industrial terms, and its correlates, such as emission or effluent growth, must be evaluated in totality. Developing a "total" view of an action's environmental impact requires consideration of:

- Structural changes in urban land-use patterns.
- The expansion and intensification of urban densities.
- Loss of open space.
- Replacement of rural form with urban form.

A major concern, however, is that growth may not be the consequence of a single project, and identifying an individual action's contribution to an overall larger pattern can prove difficult. Additionally, changes in the environment can be generated by actions that do not require environmental review. Actions of this type have proved to be particularly vexing since little information is gathered about their influences and even less is understood about how their effects accumulate to produce a change in the environment.

Cumulative impact assessment

The traditional approach to EIA tends to concentrate on the identification and evaluation of a single proposed action on its environmental setting. Comprehensive analysis of the impacts of multiple human developments on the environment has been limited. According to the language of EIA, the term "cumulative impact" is used to refer to a "holistic" approach to environmental analysis. In fact, NEPA only indirectly addresses the concept of cumulative impact in its reference to the interrelations of all environmental components. The Council on Environmental Quality addressed the concept of a cumulative impact in its 1978 guidelines. In these guidelines CEQ defined the concept of a cumulative impact as the incremental impact of multiple present and future actions with individually minor, but collectively significant, effects taking place over a period of time. Thus, cumulative impact recognizes that environmental change can be produced not only by a single action, but by the total effect of multiple land uses and developments, including their interrelationships. The important distinction made is that impacts result as a consequence of the incremental impact of an action when added to future actions as well as the product of individually minor, but collectively significant, actions that were never subject to detailed EIA review. The concept also carries the implication that the total effect of separate actions on the environment may be substantively different than the simple summation of single project effects on the landscape.

In practice, cumulative impact assessment requires the analyst to step away from the single-project focus common to EIA and adopt a wide-angle view of the scale and pace of actions occurring within the planning region. The importance of this wide-angle view has been underscored by Hamann (1984). In this study, Hamann reported on the destruction of 4,000 ha of Florida wetlands over an 18-month period. The significance of this report was that the loss of wetlands resulted from projects that were all legally permitted by state and federal agencies. Although each project, when examined individually, did not pose a significant environmental threat, taken together they resulted in the substantial loss of a critical habitat.

The objective of cumulative impact assessment is to develop approaches capable of recognizing that although individual actions may have insignificant effects, in the aggregate they become significant. Several qualities of human actions become central to this type of assessment, particularly actions that are repetitive, aggregative, continuous, or time-delayed. Identifying actions that fall into one of these categories begins the process; however, different causes may exist that generate the cumulative effect. For example, if a given action is repetitive, its effects will be incremental. If other actions are occurring concurrently or planned in the future, they will collectively affect a given environmental attribute. Finally, some effects may be discernible after a time-delay, suggesting that effects may take years to accumulate to a measurable level. Cumulative impacts are therefore synergistic. This quality must be emphasized since it clearly separates cumulative impacts from primary and secondary impacts associated with a single action.

The general principle of synergism points to the realization that the whole is more than the sum of its parts in magnitude, severity, intensity, and complexity. Synergism implies interaction, combination, and new patterning that remains a challenge to assess. Although progress toward the development of methods to guide cumulative impact assessment has been slow, four elements have been noted that are critical to an analysis:

1 The nature of the inducing action.
2 The scale or extent of ongoing transformations.
3 The rate and timing of change.
4 The characteristics of the physical setting in which the actions are taking place.

A typology of cumulative effects has been derived that provides a structure for their investigation (Cocklin et al., 1992). From this classification system, 6 central characteristics can be noted to connect an action to its relationship to cumulative change (Spaling & Smit, 1993):

- **Time crowding** – characterized by frequent and repetitive environmental changes that

Table 10.3 Forecasting methods to support cumulative impact assessment.

Method	Example
Intuitive and holistic methods	Conjecture Brainstorming Delphi techniques Heuristic programming
Scenarios and metaphors	Growth metaphors Historical analogies Scenarios Alternative futures
Extrapolation	Social trend analysis Monitoring Time-series analysis Economic forecasting models
Simulation and modeling	Structural modeling Dynamic modeling Simulation Systems analysis
Decision trees	Judgment theory Relevance trees Contextual mapping

cause the temporal capacity of an environmental medium to be exceeded.
- **Space crowdings** – resulting from a high spatial density of environmental change that alters a region's spatial pattern or processes.
- **Compoundings or synergisms** – occurring when two of more environmental changes contribute to another environmental change.
- **Time lags** – where there are delays between exposure to a perturbation and response.
- **Space lags** – where environmental changes appear some distance from their source.
- **Thresholds** – exceeding critical levels that cause disruptions to environmental processes that fundamentally alter system behavior.

From these analytical foci, the forecasting techniques needed to assist with the cumulative assessment problem are those that can compile information, direct the application of expert judgment, and track trends and deviation in environmental processes. The general forecasting methods available are summarized in Table 10.3.

The issues surrounding cumulative impact assessment and the question of growth-inducing effect shows that every analysis undertaken within the context of EIA has an implied temporal component. However, in most instances, the time horizon over which the EIA extends is never fully defined. The temporal ambiguity of EIA is even seen in NEPA and the CEQ revisions, where the analyst is asked to consider the "long term" but no specific time reference is provided. As human actions pose the possibility of carrying effects well beyond the time horizons of most environmental and comprehensive plans, a mechanism is needed that articulates time more strongly and uses time to evaluate how plans and actions will perform well into the future.

Time and the long term

Questions regarding the irreversibility of human impact and the long-term viability of environmental systems threatened by human activities have been difficult to resolve (Lein, 1992). Characterizing human impact requires a set of variables as diverse and complex as those governing the environmental system under consideration. Complicating matters is the high level of uncertainty that surrounds environmental processes and the form, diffusion, and persistence of human impact. The speed with which modern technology is developing assures that decisions made by society will increasingly risk affecting environmental systems well beyond the planning horizons of present-day policy-making institutions (Malone, 1990). This observations is evidenced in the controversies surrounding recent concerns over global warming, biodiversity, nuclear waste disposal, hazardous waste management, and habitat destruction. Such issues have tremendous implications for future societies that may be asked to adapt to an environmental systems that has shifted to a new equilibrium. To meet this challenge, new methodologies must be developed to evaluate the impact of human actions that may have environmental consequences that extend for periods of 100 to 10,000 years or more.

Environmental impact assessment has been the principal method for understanding human impacts on the environment within a structured decision-making framework. However, EIA traditionally emphasizes short-term impacts. The short-term nature of EIA was identified in a study conducted by Coates and Coates (1989). In this study the authors concluded that the existing concepts and methods that guide impact assessment are inadequate when effects over time periods exceeding 50 to 100 years are considered. Although forecasting is an important aspect of the process, the primary role of EIA is to guide decision-makers in making an informed trade-off among conflicting features of a proposed action (Culhane et al., 1989). As a result, EIA tends to be deficient in providing decision-makers with defensible forecasts when human actions and the environment are viewed in concert over the long term (Duinker & Baskerville, 1986).

In light of recent controversies regarding the nature of human impact on regional and global systems, and the increased potential modern society displays for engendering irreversible changes in vital Earth-system processes, a model is sought that can forecast the long-term nature of human action and constructions, communicating their implications at some future state of the environment. The emerging technique of environmental performance assessment holds promise in this regard, although tractable methods for conducting such studies remain in the developmental stages.

The environmental performance assessment

The concept of environmental performance assessment has been most fully developed within the context of high-level nuclear waste repository siting in complex geologic settings (Malone, 1990; Lemons & Malone, 1988). Using nuclear waste disposal as a backdrop, a performance assessment provides a quantified description of a system's current behavior, its expected future behavior, and the acceptability of that behavior when compared to a set of standards that specify the degree of safety required in the system over time. Conducting the performance assessment entails a detailed analysis and documentation of the processes, events, and uncertainties that could act to destabilize a nuclear waste repository site; it also entails identifying the potential consequences of one or more destabilizing events and their likelihood of occurrence. Given this information, the ability of a site to "perform" in accordance with a set of safety (performance) standards can be evaluated before the site is committed to a use with irreversible cross-generational consequences.

With respect to the repository siting example, the spent nuclear fuel must remain isolated from the biosphere for a minimum of 10,000 years, and a site must remain geologically stable while environmental processes and potential human interference act on the location over time. Clearly, this is a decision task well beyond the scope of EIA. Considering the magnitude and risks associated with the siting problem, the question is not only whether a particular site, given its environmental characteristics, will perform, but whether the approach taken to evaluate its future state is adequate. To address these issues, an environmental performance assessment must focus on four critical aspects of a proposed action:

1 The risks associated with the environmental components that describe the site.
2 The interaction between environmental components as presently understood through the application of predictive models across the selected time horizon.
3 The nature and significance of uncertainty.
4 The impact and ramifications of a failure in performance.

Beyond the high-level waste disposal problem, similar situations can be described where prolonged exposure to human activities is threatening to drive environmental systems to a new state. In these situations environmental performance assessment can be expanded to include not merely an engineering system or construction, but also policy decisions whose implications may carry extended environmental consequences. The advantage of the performance assessment in these applications is its ability to focus on the mitigation measures available to reduce adverse consequences and to investigate whether mitigation

will keep potential risks below environmental thresholds at extended time scales. In addition, the uncertainties associated with the action can be identified and environmental trends attributable to the action can be projected to explore various "failure" scenarios.

Conducting a performance assessment

Based on our previous discussion, environmental performance explains the quantified description of a system's present behavior, its expected future behavior, and the acceptability of that behavior in relation to a set of "performance" standards that will reduce the risk of adverse environmental consequences. The environmental performance assessment, therefore, involves two interrelated tasks:

1 The analysis and documentation of the processes, events, and uncertainties that influence long-term system behavior.
2 The identification of the potential consequences of one or more system events and their likelihood.

To be effective the performance assessment must provide information concerning the total reliability of the environmental components that characterize the site and their interaction. Brandstetter and Buxton (1989) have summarized the procedures followed to derive this information. In general, assessment consists of a series of iterative steps that begin with the collection of facts and data characterizing the planned construction and culminate in an estimate of confidence in the system's performance though sensitivity and uncertainty analysis. The basic outline for the method of long-term environmental impact forecasting involves:

1 Collecting facts and data characterizing the planned construction (action).
2 Developing conceptual and analytical models describing the relevant human–environmental interactions related to the construction.
3 Identifying the events (natural and human induced) that could trigger a failure in the system's performance.
4 Developing a series of scenarios that explain

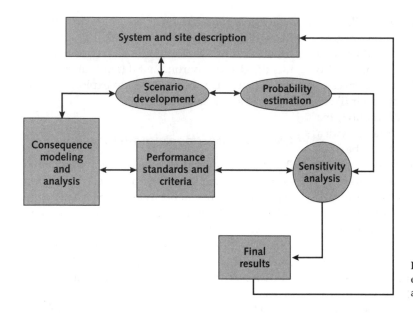

Fig. 10.3 Features of the environmental performance assessment process.

the event identified in (3) above, and detailing their consequences.

5 Selecting a set of credible scenarios and modeling their behavior over the analytic time horizon.

6 Evaluating the results of the modeling effort.

7 Reviewing and adjusting the construction as necessary to ensure its long-term performance effectiveness (safety).

The basic procedures presented above can be illustrated to reveal the overall flow of information through the process (Fig. 10.3). Perhaps the most critical stages in performance assessment involve developing conceptual and analytical models of human–environment interactions, identifying triggering events that could lead to a system failure, and detailing the consequences that could result from such failures. Processing through these stages leads to the selection and analysis of scenarios that provide decision-makers with a set of probability estimates surrounding the consequence of a given event/failure sequence.

Since the future is obviously difficult to understand, performance assessment provides a crystal ball that promotes a model-focused analysis that can be employed to examine the implications of technologies and actions that are capable of affecting future generations. However, while it is a promising tool, there are several unresolved issues that must be addressed in order to improve its wider application:

- Methods are needed for quantifying uncertainties associated with parameters characterizing natural systems.
- A rigorous objective approach is needed to assist with developing scenarios.
- Methods are needed for estimating probabilities and for measuring uncertainty.
- Systematic approaches are needed to guide the use of expert judgment.
- Validated computation models are needed to direct quantitative and predictive analysis of complex natural processes.
- Better understanding is needed of the uncertainties that influence the behavior of natural systems over long time periods.

As these issued are addressed, the planner's ability to look at change over extended time horizons will be enhanced and the challenges of the future placed within our grasp.

The continuing challenge

Environmental impact assessment and the emerging technique of environmental performance assessment identify methods that personify the proactive nature of environmental planning and the need to integrate "future thinking" into all aspects of the planning process. As the needs of modern society increase demand for land and other resources, maintaining balance between the human world and the environmental systems on which that world depends will remain a challenge. Meeting this challenge will require wisdom and creativity together with some fundamental changes in the way we look at our world. Change, however, is something difficult to conceptualize and even more difficult to accept; yet the need to plan for change is greater now than ever before.

As we consider the implications of social and ecological perturbations resulting from shifting lifestyles and technical advancements over the past 25 years, we must begin to consciously and deliberately continue our pursuit of a new system of physical order, bridging the local with the global, and taking full advantage of the data and information that has been acquired to guide us. The missing ingredient is a systematic and considered approach to the issues that evidence change. Fortunately the natural environment offers us numerous examples and models we can follow. Thus, by beginning simply, societal awareness can be enhanced and may show us that while complex, the environment is not complicated. Solutions are possible; we simply need to look for them.

For the last decade, it has been convenient to view environmental issues as "global" – distancing ourselves from the problems and placing responsibility for them on an abstraction removed from our day-to-day reality. In actuality few environmental problems are global or national. While their total effect may be reflected at these scales, their origins are closer to home and ultimately the product of individual decisions made in places most tend to ignore. Urban patterns expand because we live in dispersed settlements that are removed from the disamenities of urban life. Water supplies become scarce because we enjoy using this resource in wide and varied ways. Farmland disappears because we fail to recognize that these are the places our food comes from. We build on steep slopes because we like the view. We consume natural habitat because we have our own habitat needs. These local decisions color the environmental problem, direct how the planning process responds, and influence what our future will look like. The forgotten element in all this is that it is our future, and each of us helps to shape and guide the changes it will reveal. To be effective, each of us must contribute to a greater understanding of the environment, learning from the landscapes we inhabit and applying those lessons to select alternatives that will build toward a sustainable future. Here, we can recall a simple paragraph taken from Aldo Leopold's *A Sand County Almanac* written some 50 years ago:

The last word in ignorance is the man who says of an animal or plant: "What good is it?" If the land mechanisms as a whole is good, then every part is good, whether we understand it or not. If the biota in the course of time, has built something we like, but do not understand, then who but a fool would discard seemingly useless parts? To keep every cog and wheel is the first precaution of intelligent tinkering.

Intelligent tinkering begins with a plan and the synthesis of information regarding the environment, its opportunities, constraints, and risks that will communicate ecologically sound principles to decision-makers and recommend only that course of action that makes a difference in a manner that fits logically and naturally into our lives. This is the challenge, and it has been the aim of this text to provide directions to the point where this journey begins.

Summary

The proactive nature of environmental planning directs us to examine the characteristics of a proposed plan or project before committing to an irreversible outcome that may carry serious environmental consequences. Consideration of the qualities of a plan or proposed action superimposed onto the natural system and projected into an uncertain future is the subject of this chapter.

Two important recipes for assessing the future were described: environmental impact assessment and environmental performance assessment. As defined in the US NEPA, EIA requires a detailed analysis of all federal actions that significantly affect the quality of the human environment. Approaches followed when conducting such an analysis were explored. Yet EIA alone does not provide sufficient guidance, particularly when time scales exceed 100 years. When confronted with time-extensive problems, environmental performance assessment provides a methodology for evaluating long-term risk. Although it is still an emerging technique, along with EIA it is an important tool to reduce uncertainty and better understand the potential ramifications of a human decision. Yet the challenges to develop sustainable plans and reduce the entropic effects of human actions remains great.

Focusing questions

Explain the importance of disclosure in EIA.

Outline the fundamental activities that define the method of EIA.

Compare and contrast growth-inducing impacts with cumulative impacts; how can they be identified?

Explain the basic steps involved in long-term impact forecasting.

How might techniques such as performance assessment and EIA reshape environmental decision-making?

References

Aageenbrug, R. 1991. *A Critique of GIS in Geographical Information Systems: Principles and Applications*, ed. D. Maguire, London: Longman.

Adams, J. 1974. *Conceptual Blockbusting*, San Francisco: W. H. Freeman.

Ahrens, C. 1991. *Meteorology Today*, St. Paul, MN: West Publishing Co.

Allen, W. 1984. The Palmer Drought Severity Index: Limitations and Assumptions, *Journal of Climate and Applied Meteorology* 23, 1100–09.

Alonso, W. 1964. *Location and Land Use: Toward a General Theory of Land Rent*, Cambridge, MA: Harvard University Press.

Alterman, R. and Hill, M. 1978. Implementation of Urban Land Use Plans, *Journal of the American Institute of Planners* 33(3), 274–85.

Anderson, L. 1995. *Guidelines for Preparing Urban Plans*, Chicago, IL: American Planning Association.

Anderson, P. 1980. *Regional Landscape Analysis*, Reston, VA: Environmental Design Press.

Andruss, V., Plant, C., Plant, J., and Wright, E. (eds.) 1990. *Home: A Biogressional Reader*, Philadelphia: New Society Publishers.

Antrop, M. 1998. Landscape Change: Plan or Chaos, *Landscape and Urban Planning* 41, 155–61.

Archibusi, F. 1997. *The Ecological City and the City Effect*, Aldershot, UK: Ashgate Publishing.

Attwell, K. 1991. *An Outline on Theories, Practice and Results from Working with Urban Ecology in Denmark*. European Center for the Environment Research Colloquium, Copenhagen.

Bailey, J. 1997. Environmental Impact Assessment and Management: An Underexplored Relationship, *Environmental Management* 21(3), 317–27.

Bailey, R. 1988. Problems with Using Overlay Maps for Planning and Tier Implications for Geographic Information Systems, *Environmental Management* 12(1), 11–17.

Baldwin, J. 1985. *Environmental Planning and Management*, Boulder, CO: Westview Press.

Barber, D. 1981. *Citizen Participation in American Communities*, Dubuque, IA: Kendall/Hunt.

Barlowe, R. 1972. *Land Resource Economics*, Englewood Cliffs, NJ: Prentice-Hall.

Barrett, B. F. and Therivel, R. 1991. *Environmental Policy and Impact Assessment in Japan*, New York: Routledge, Chapman and Hall.

Bartlett, R. and Malone, C. 1993. Guidance for Environmental Impact Assessment, *The Environmental Professional* 15(1), 1–5.

Batty, M. and Xie, Y. 1994. From Cells to Cities, *Environment and Planning B: Planning and Design* 24, pp. 175–92.

Batty, M. and Xie, Y. 1997. Possible Urban Automata, *Environment and Planning B: Planning and Design* 24, 175–92.

Bayne, P. 1995. Generating Alternatives: A Neglected Dimension in Planning Theory, *Town and Planning Review* 66(3), 303–20.

Beanlands, G. 1989. Scoping Methods and Baseline Studies in EIA, in *Environmental Impact Assessment: Theory and Practice*, ed. P. Wathern, London: Unwin Hyman.

Beatley, T. 1994. *Ethical Land Use: Principles of Policy and Planning*, Baltimore, MD: John Hopkins University Press.

Beatley, T. 1995. Planning and Sustainability: The Elements of a New Paradigm, *Journal of Planning Literature* 9(4), 383–95.

Bell, E. 1974. Markor Analysis of Land Use Change, *Journal of Socio-Economic Planning Science* 8, 311–16.

Bennet, R. and Chorely, R. 1978. *Environmental Systems: Philosophy, Analysis and Control*, London: Methuen.

Berg, P. and Dasmann, R. 1978. Reinhabiting California, in Berg, P. (ed.), *Reinhabiting a Separate Country*, San Francisco: Planet Drum Foundation.

Berry, J. 1995. *Spatial Reasoning for Effective GIS*, Boulder, CO: GIS World.

Betters, D. and Rubingh, J. 1978. Suitability Analysis and Wildland Classification, *Journal of Environmental Management* 7, 59–72.

Bishop, A. et al. 1974. *Carrying Capacity in Regional Environmental Management*, Washington, DC: Government Printing Office.

Bisset, R. 1988. Developments in EIA Methods, in *Environmental Impact Assessment: Theory and Practice*, ed. P. Wathern, London: Unwin Hyman.

Blaike, P., Cannon, T., Davis, I., and Wisner, B. 1994. *At Risk: Natural Hazards, People's Vulnerability and Disasters*, New York: Routledge.

Bonano, E. et al. 1989. Treatment of Uncertainty in the Performance Assessment of Geologic High-Level Radioactive Waste Repositories, *Journal of the International Association for Mathematical Geology* 20, 543–65.

Bonham-Carter, G. 1994. *Geographic Information Systems for Geoscientists*, Tarryton, NJ: Pergamon/Elsevier Science Publications.

Bourne, L. 1971. Physical Adjustment Processes and Land Use Succession, *Economic Geography*, 47, 1–15.

Brandstetter, A. and Buxton, B. 1989. The Role of Geostatistical, Sensitivity, and Uncertainty Analysis in Performance Assessment, in *Geostatistical, Sensitivity, and Uncertainty Methods for Ground-Water Flow and Radionuclide Transport Modeling*, eds. B. Buxton et al., Battelle Press, Columbus, OH, pp. 89–110.

Brennan, A. 1995. *The Ethics of the Environment*, Aldershot, UK: Dartmouth.

Briassoulis, H. 1989. Theoretical Orientations in Environmental Planning: An Inquiry into Alternative Approaches, *Environmental Management* 13(4), 381–92.

Brinkman, R. and Smyth, A. 1973. Land Evaluation for Rural Purposes Publication No. 17. International Institute for Land Reclamation, Wageningen, the Netherlands.

Bryson, J. 1990. Influence of Context and Process on Project Planning Success. *Journal of Planning Education and Research* 9(3), 183–95.

Buchholz, R. 1993. *Principles of Environmental Management*, Englewood Cliffs, NJ: Prentice-Hall.

Bullard, Robert D. 1996. Environmental Justice: It's More than Waste Facility Siting. *Social Science Quarterly* 77(3): 493–9.

Burby, R. 1998. *Cooperating with Nature*, Washington, DC: Joseph Henry Press.

Burrough, P. 1986. *Principles of Geographical Information Systems for Local Resource Assessment*. Oxford: Clarendon Press.

Burrough, P. and Frank, A. 1996. *Geographic Objects with Indeterminate Boundaries*, London: Taylor and Francis.

Burrough, P. and McDonnell, R. 1998. *Principles of Geographical Information Systems*, Oxford: Oxford University Press.

Bush, M. 2000. *Ecology for a Changing Planet*, Upper Saddle River, NJ: Prentice-Hall.

Cairns, J. 1991. The Need for Integrated Environmental Systems Management, in Cairns, J. and Crawford, T. (eds.), *Integrated Environmental Management*, Boca Raton, FL: Lewis Publishers.

Caldwell, L. 1997. Implementing NEPA: A Non-Technical Political Task, in *Environmental Policy and NEPA*, ed. R. Clark, Boca Raton, FL: St. Lucie Press.

Callicott, J. and Mumford, K. 1997. Ecological Sustainability as a Conservation Concept, *Conservation Biology* 11(1), 32–40.

Campbell, H. 1994. How Effective is GIS in Practice? *International Journal of Geographical Information Systems* 8, 309–25.

Canter, L. 1993. Pragmatic Suggestions for incorporating Risk Assessment Principles in EIA studies, *The Environmental Professional* 15, 125–38.

Canter, L. 1996. *Environmental Impact Assessment*, New York: McGraw-Hill.

Cartwright, T. 1993. *Modeling the World in a Spreadsheet*, Baltimore, MD: Johns Hopkins University Press.

Castillon, D. 1996. *Conservation of Natural Resources*, Madison, WI: Brown and Benchmark Publishers.

Catanese, A. and Snyder, J. 1988. *Urban Planning*, New York: McGraw-Hill.

Catton, W. 1986. Carrying Capacity and the Limits to Freedom, paper prepared for Social Ecology Session World Congress of Sociology, New York.

Cecchini, A. and Viola, F. 1992. *Ficties: A Simulation for the Creation of Cities*, International Seminar on Cellular Automata, Venice, Italy.

Chadwick, G. 1974. *A Systems View of Planning*, Oxford: Pergamon Press.

Chrisman, N. 1997. *Exploring Geographic Information Systems*, New York: Wiley and Sons.

Churchman, C. 1967. Wicked Problems, *Management Science* 14(4), 141–2.

Cifuentes, P. et al. 1995. Techniques for Multivariate Analysis, in *Quantitative Techniques in Landscape Planning*, ed. E. Falero, Boca Raton, FL: CRC Lewis Publishers.

Clakins, H. 1979. The Planning Monitor: An Accountability Theory of Plan Evaluation, *Environment and Planning A* 11(7), 745–58.

Clark, R. 1997. NEPA: The Rational Approach to Change, in *Environmental Policy and NEPA*, ed. R. Clark, Boca Raton, FL: St. Lucie Press.

Clarke, K., Brass, J., and Riggan, P. 1994. A Cellular Automation Model of Wildfire Propagation and Extinction, *Photogrammetric Engineering and Remote Sensing* 60, 1355–67.

Clarke, K., Hoppen, S., and Gaydos, L. 1997. A Self-Modifying Cellular Automation Model of Historical Urbanization in the San Francisco Bay Areas, *Environment and Planning B: Planning and Design* 24, 247–61.

Coates, V. and Coates, J. 1989. Making Technology Assessment an Effective Tool to Influence Policy, in *Policy Through Impact Assessment*, ed. R. Bartlett, New York: Greenwood Press.

Cocklin, C. et al. 1992. Notes on Cumulative Environmental Change I: Concepts and Issues, *Journal of Environmental Management* 35, 31–49.

Cook, S. 1972. *Prehistoric Demography*, Modular Publication 16, New York: Addison-Wesley.

Cooke, R. and Doorkamp, J. 1990. *Geomorphology in Environmental Management*, Oxford: Clarendon Press.

Cowen, D. 1989. GIS versus CAD versus DBMS. *Photogrammetric Engineering and Remote Sensing* 54, 1551–4.

Culhane, J. et al. 1989. *Forecasts and Environmental Decision-Making*, Boulder, CO: Westview Press.

Davidson, D. 1992. *The Evaluation of Land Resources*, Essex, UK: Longman Group.

Davis, J. 1986. *Statistical and Data Analysis in Geology*, New York: Wiley and Sons.

Davis, M. 1988. *Applied Decision Support*, New York: Prentice-Hall.

Day, D. 1997. Citizen Participation in the Planning Process, *Journal of Planning Literature* 11(3), 421–34.

Dear, M. 1992. Understanding the NIMBY Syndrome, *Journal of the American Planning Association* 58(3), 288–300.

Debons, A., Horne, E., and Cronenweth, S. 1988. *Information Science: An Integrated View*. Boston: G. K. Hall and Co.

DeJongh, P. 1988. Uncertainty in EIA, in *Environmental Impact Assessment: Theory and Practice*, ed. P. Wathern, London: Unwin Hyman.

Dent, D. and Young, A. 1981. *Soil Survey and Land Evaluation*, London: George Allen and Unwin.

DeSario, J. and Langton, S. 1987. *Citizen Participation in Public Decision Making*, Westport, CT: Greenwood.

Dewdney, A. 1990. The Cellular Automata Programs that Create Wireworld, *Scientific American* 262(1), 146–9.

Diaz, H. 1983. Some Aspects of Major Dry and Wet Periods in the Contiguous United States, 1895–1981, *Journal of Climate and Applied Meteorology* 22, 3–16.

Dickerson, W. and Montgomery, J. 1993. Substantive Scientific and Technical Guidance for NEPA Analysis, *The Environmental Professional* 15(1), 7–11.

Dorney, R. 1989. *The Professional Practice of Environmental Management*, New York: Springer-Verlag.

Douglas, I. 1983. *The Urban Environment*, London: Edward Arnold.

Downey, L. 1998. Environmental Justice: Is Race or Income a Better Predictor? *Social Science Quarterly* 79(4): 766–78.

Dramstad, W. et al. 1996. *Landscape Ecology Principles in Landscape Architecture and Land use Planning*. Washington, DC: Island Press.

Drought Information Center. 1999. *Understanding and Defining Drought*, National Drought Mitigation Center, University of Nebraska–Lincoln, http://enso.unl.edu/ndmc/enigma/def2.htm

Duchhar, I. 1989. *Manual on Environmental and Urban Development*, Ministry of Local Government and Physical Planning, Nairobi, Kenya.

Duinker, P. and Baskerville, G. 1986. A Systematic Approach to Forecasting in Environmental Impact Assessment, *Journal of Environmental Management* 23: 271–90.

Dunn, R., Harrison, A., and White, J. 1990. Positional Accuracy and Measurement Error in Digital Databases of Land Use, *International Journal of Geographical Information Systems* 4, 385–98.

Dunne, T. and Leopold, L. 1978. *Water in Environmental Planning*, San Francisco: W. H. Freeman.

Eastman, R. et al. 1993. *GIS and Decision Making*. Explorations in GIS Technology 4, Geneva: Unitar.

Engelen, G. et al. 1996. Numerical Modeling of Small Island Socioeconomics, in G. Maul (ed.), *Small Islands: Marine Science and Sustainable Development*, Washington, DC: American Geophysical Union, 437–63.

Erickson, P. 1994. *A Practical Guide to Environmental Impact Assessment*, San Diego, CA: Academic Press.

Evans, J. and Olson, D. 1998. *Introduction to Simulation and Risk Analysis*, Upper Saddle River, NJ: Prentice-Hall.

Fabos, J. 1979. *Planning the Total Landscape: A Guide to Intelligent Land Use*. Boulder, CO: Westview Press.

Fabos, J. and Caswell, S. 1977. Composite Landscape Assessment: Assessment Procedures for Special Resources, Hazards, and Development Suitability (METLAND), research Bulletin 637, Massachusetts Agricultural Experiment Station, Amherst, MA.

Faust, N. 1995. The Virtual Reality of GIS, *Environmental and Planning B: Planning and Design* 22, 257–68.

Feldt, A. 1988. Planning Theory, in Cantanese, A. and Snyder, J. (eds.), *Urban Planning*, New York: McGraw-Hill.

Finn, J. 1993. Use of the Average Mutual Information Index in Evaluating Classification Error, *International Journal of Geographical Information Systems* 7, 349–66.

Fischer, F. 1993. Citizen Participation and the Democratization of Policy Expertise. *Policy Science* 26, 165–87.

Flawn, P. 1970. *Environmental Geology*, New York: Harper and Rowe.

Flemming, D. 1991. Expert Judgment and High-Level Nuclear Waste Management, *Policy Studies Review* 11, 311–31.

Ford, A. 1999. *Modeling the Environment*, Washington, DC: Island Press.

Foreman, R. 1998. *Land Mosaics: The Ecology of Landscapes and Regions*, Cambridge: Cambridge University Press.

Foreman, R. and Gordon, M. 1986. *Landscape Ecology*, New York: Wiley and Sons.

Forester, J. 1966. *Urban Dynamics*, Cambridge, MA: MIT Press.

Foth, H. 1978. *Fundamentals of Soil Science*, New York: Wiley and Sons.

Fowles, J. 1978. *Handbook of Futures Research*, Westport, CT: Greenwood Press.

Frankena, W. 1973, *Ethics*, Englewood Cliffs, NJ: Prentice-Hall.

Gardner, M. 1970. The Fantastic Combinations at John Conway's Game of Life, *Scientific American* 223(4), 120–3.

George, R. 1994. Formulating the Right Planning Problem, *Journal of Planning Literature* 8(3), 240–59.

Geraghty, P. 1992. Environmental Assessment and the Application of an Expert Systems Approach, *Town and Planning Review* 63, 123–42.

Godet, M. and Roubelat, F. 1996. Creating the Future: The Use and Misuse of Scenarios, *Long Range Planning* 29(2), 164–71.

Godschalk, D., Kaiser, E., and Berke, P. 1998. Integrating Hazard Mitigation and Local Land Use Planning, in *Cooperating with Nature*, ed. R. Burby, Washington, DC: Joseph Henry Press.

Gonzalez-Alonso, S. 1995. Using the Inventory: Models in Landscape Planning, in *Quantitative Techniques in Landscape Planning*, ed. E. Falero, Boca Raton, FL: Lewis Publishers.

Gordon, M. and Foreman, R. 1983. Landscape Modification and Changing Ecological Characteristics, in *Disturbance and Ecosystems*, ed. H. Mooney, New York: Springer-Verlag.

Gordon, S. 1985. *Computer Models in Environmental Planning*, New York: Van Nostrand Reinhold.

Grazulis, T. 1993. *Significant Tornadoes 1680–1991: A Chronology and Analysis of Events*, St. Johnsbury, VT: The Tornado Project.

Griggs, G. and Gilchrist, J. 1977. *The Earth and Land Use Planning*, North Scituate, MA: Duxbury Press.

Hall, A. and Fagen, R. 1959. Definition of a System, *General Systems Yearbook* 1, 18–28.

Hamann, R. 1984. Legal Issues in Water Resource Management, in *Water Resources, Atlas of Florida*, eds. E. Fernald et al., Tallahassee, Florida Institute for Science and Public Affairs.

Han, S. and Kim, T. 1989. Can Expert Systems Help with Planning? *Journal of the American Planning Association* 34: 296–308.

Handmer, J. 1990. Approaches to Flood Damage Assessment, in *Annual Flood: Evaluation for a Computerized Flood Loss Assessment*, ed. N. J. Ericksen Wellington, NZ: National Water and Soil Conservation Authority.

Hanna, S. 1995. Efficiencies of User Participation in Natural Resource Management, in *Property Rights and the Environment*, ed. S. Hanna, Washington, DC: The World Bank.

Hannon, B. and Ruth, M. 1994. *Dynamic Modeling*. New York, Springer-Verlag.

Harbaugh, J. and Bonham-Carter, G. 1981. *Computer Simulation in Geology*, Malabar, FL: Krieger Publishing.

Hardisty, J., Taylor, D., and Metcalfe, S. 1993. *Computerized Environmental Modeling*, Chichester, UK: Wiley and Sons.

Hargrove, E. 1985. The Role of Rules in Ethical Decision Making, *Inquiry* 28, 3–42.

Hargrove, E. 1989. *Fundamentals of Environmental Ethics*, Englewood Cliffs, NJ: Prentice-Hall.

Harris, B. and Batty, M. 1993. Locational Models, Geographical Information Systems and Planning Support Systems, *Journal of Planning Education and Research* 12, 184–98.

Hassan, F. 1982. Demographic Archaeology, in *Advances in Archaeological Methods and Theory*, ed. M. Schiffer, New York: Academic Press.

Hayden, B. 1975. The Carrying Capacity Dilemma, *American Antiquity* 40(2), 11–21.

Henderson-Sellers, A. and Robinson, P. 1986. *Contemporary Climatology*, London: Longman Scientific.

Hersperger, A. 1994. Landscape Ecology and Its Potential Application to Planning, *Journal of Planning Literature* 9(1), 14–29.

Hidore, J. and Oliver, J. 1993. *Climatology: An Atmospheric Science*, New York: Macmillan Publishing.

Hill, K. 1978. Trend Extrapolation, in *Handbook of Futures Research*, ed. J. Fowles, Westport, CT: Greenwood Press.

Hillier, F. and Lieberman, G. 1967. Introduction to Operations Research, San Francisco: Holden-Dag.

Holmberg, S. 1994. Geoinformatics for Urban and Regional Planning. *Environment and Planning B: Planning and Design* 21: 5–19.

Honachefsky, W. 2000. *Ecologically Based Municipal Land Use Planning*, Boca Raton, FL: Lewis Publishers.

Hopkins, L. 1977. Methods for Generating Land Suitability Maps. *Journal of the American Institute of Planners* 43, 386–400.

Hough, M. 1989. *City Form and Natural Process*, London: Routledge.

Hough, M. 1995. *Cities and Natural Process*. London: Routledge.

Huggett, R. 1980. *Systems Analysis in Geography*, London: Clarendon Press.

Hushon, J. 1990. *Expert Systems for Environmental Applications*, ACS Symposium Series 431, Washington, DC: World Bank Publications.

Huxhold, W. 1991. *An Introduction to Urban Geographic Information Systems*, Oxford: Oxford University Press.

Huxhold, W. and Levinsohn, A. 1994. *Managing Geographic Information System Projects*, Oxford: Oxford University Press.

Inhaber, H. 1998. *Slaying the NIMBY Dragon*, New Brunswick, NJ: Transaction Publishers.

Jacobs, H. 1995. Contemporary Environmental Philosophy and Its Challenge to Planning Theory, in *Planning Ethics*, ed. S. Hendler, New Brunswick, NJ: Center for Urban Policy Research.

Johnson, R. and DeGraff, J. 1988. *Principles of Engineering Geology*, New York: Wiley and Sons.

Kachigan, S. 1986. *Statistical Analysis*, New York: Radius Press.

Kaiser, E., Godschalk, D., and Chapin, F. 1995. *Urban Land Use Planning*, Urbana, IL: University of Illinois Press.

Kartez, J. and Lindell, M. 1987. Planning for Uncertainty: The Case of Local Disaster Planning. *Journal of the American Planning Association* 53(4), 487–98.

Katz, R. and Glantz, M. 1986. Anatomy of a Rainfall Index, *Monthly Weather Review* 134, 764–71.

Keller, E. 1988. *Environmental Geology*, Columbus, OH: Merrill Publishing.

Kelly, E. and Becker, B. 2000. *Community Planning*, Washington, DC: Island Press.

Kim, T., Wiggins, L., and Wright, J. 1990. *Expert Systems Applications to Urban Planning*, New York: Springer-Verlag.

Kirkby, S. 1996. Integrating a GIS with an Expert System to Identify and Manage Dryland Salinization, *Applied Geography* 16(4), 289–303.

Klee, G. 1999. *The Coastal Environment*, Upper Saddle River, NJ: Prentice-Hall.

Klein, M. and Methlie, L. 1990. *Expert Systems: A Decision Support Approach*, Wokingham, UK: Addison-Wesley.

Klingebiel, A. and Montgomery, P. 1961. *Land-Capability Classification*, USDA Agriculture Handbook no. 210.

Kliskey, A. 1995. The Role and Functionality of GIS as a Planning Tool in Natural Resource Management, *Computers, Environment, and Urban Systems* 19, 15–22.

Klosterman, R. 1998. The What If? Collaborative Planning Support Systems, *Environment and Planning B: Planning and Design* 26, 393–408.

Klosterman, R., Brail, R., and Bossard, E. 1993. *Spreadsheet Models for Urban and Regional Analysis*, New Brunswick, NJ: Center for Urban Policy Research.

Koplik, C., Kaplan, M., and Ross, B. 1982. The Safety of Repositories for Highly Radioactive Wastes, *Review of Modern Physics* 54, 93–117.

Ladd, J. 1973. *Ethical Relativism*, Belmont, CA: Wadsworth Publishing.

Lahde, J. 1982. *Planning for Change: A Course of Study in Ecological Planning*, New York: Teachers College Press.

Langendorf, R. 1985. Computers and Decision Making, *Journal of the American Planning Association*, 27, 422–33.

Largo, J. 2001. *Site Analysis: Linking Program and Concept in Land Planning and Design*, New York: John Wiley & Sons.

Lawrence, D. 1994. Designing and Adapting the EIA Planning Process, *The Environmental Professional* 16, 2–21.

Lawrence, D. 1997. Integrating Sustainability and Environmental Impact Assessment, *Environmental Management* 21(1), 23–42.

Lee, N. 1983. The Future Development of Environmental Impact Assessment, *Journal of Environmental Management* 14, 71–90.

Legget, R. 1973. *Cities and Geology*, New York: McGraw-Hill.

Lein, J. 1989. An Expert System Approach to Environmental Impact Assessment, *International Journal of Environmental Studies* 33, 13–27.

Lein, J. 1990. Exploring a Knowledge-Based Procedure for Developmental Suitability Analysis, *Applied Geography* 10, 171–86.

Lein, J. 1992. Expressing Environmental Risk Using Fuzzy Variables, *The Environmental Professional* 14, 257–67.

Lein, J. 1993a. Applying Expert Systems Technology to Carrying Capacity Assessment, *Journal of Environmental Mangement* 37, 63–84.

Lein, J. 1993b. Formalizing Expert Judgment in the Environmental Impact Assessment Process, *The Environmental Professional* 15, 95–102.

Lein, J. 1997. *Environmental Decision Making: An Information Technology Approach*, Malden, MA: Blackwell Science.

Lein, J. 1998. Automating the Environmental Impact Assessment Process: An Implementation Perspective, *Applied Geographic Studies* 2(1), 59–75.

Leitman, J. 1999. *Sustaining Cities: Environmental Planning and Management in Urban Design*, New York: McGraw-Hill.

Lemons, J. 1987. Environmental Science and Valves, *The Environmental Professional* 9(4), 277–8.

Lemons, J. and Malone, C. 1988. Siting America's Geologic Repository for High-Level Nuclear Waste, *Environmental Management* 13, 435–41.

Leopold, L. et al. 1971. *A Procedure for Evaluating Environmental Impact*, Circular 645, US Geological Survey, Washington, DC.

Leung, H. 1989. *Land Use Planning Made Plain*, Kingston, Ont.: Ronald P. Frye and Co.

Levy, J. 1997. *Contemporary Urban Planning*, Upper Saddle River, NJ: Prentice-Hall.

Linstone, H. and Turoff, M. 1975. *The Delphi Method: Techniques and Applications*, Reading, MA: Addison-Wesley.

Lugo, A. 1978. Stress and Ecosystems, in *Energy and Environemntal Stress in Aquatic Ecosystems*, ed. J. Thorp, Washington, DC: US Department of Energy.

MacKinnon, B. 1995. *Ethics, Theory and Contemporary Issues*, Belmont, CA: Wadsworth.

Maguire, D. 1991. An Overview and Definition of GIS, in *Geographical Information Systems: Principles and Applications*, ed. D. Maguire, New York: Wiley and Sons.

Malik, M. and Bartlett, R. 1993. Formal Guidance for the Use of Science in EIA, *The Environmental Professional* 15(1), 34–45.

Malone, C. 1990. Environmental Performance Assessment: A Case Study of an Emerging Methodology, *Journal of Environmental Systems* 19(2), 171–84.

Margalef, R. 1970. *Perspectives on Ecological Theory*, Chicago: University of Chicago Press.

Margerun, R. 1997. Integrated Approaches to Environmental Planning and Management, *Journal of Planning Literature* 11(4), 459–75.

Marriot, B. 1997. *Environmental Impact Assessment*, New York: McGraw-Hill.

Marsh, W. 1997. *Landscape Planning*, New York: Wiley and Sons.

Mason, D. 1994. Scenario-Based Planning: Decision Model for the Learning Organization, *Planning Review*, March–April.

McAllister, D. 1986. *Evaluation in Environmental Planning*, Cambridge, MA: MIT Press.

McCloy, K. 1995. *Resource Management Information Systems: Process and Practice*, New York: Taylor and Francis.

McDowell, B. 1986. Approaches to Planning, in *The Practice of State and Regional Planning*, eds. F. So et al., Chicago: American Planning Association.

McEvoy, J. and Dietz, T. 1977. *Handbook for Environmental Planning: The Social Consequences of Environmental Change*, New York: Wiley.

McHarg, I. 1969. *Design with Nature*, Garden City, NJ: Natural History Press.

McRae, S. and Burnham, C. 1981. *Land Evaluation*, Oxford: Clarendon Press.

Milkov, F. 1976. A Defense of Anthropogenic Landscape Science, *Soviet Geography* 17, 190–6.

Miller, D. and De Roo, G. 1997. *Urban Environmental Planning*, Aldershot, UK: Avebury.

Miller, G. 1998. *Living in the Environment*, Belmont, CA: Wadsworth Publishing.

Mitchell, B. 1989. *Geography and Resource Analysis*, Essex, UK: Longman.

Moffat, I. 1990. The Potentialities and Problems Associated with Applying Information Technology to Environmental Management, *Journal of Environmental Mangement* 30, 209–20.

Mohai, P. and Bryant, B. 1992. *Race and the Incidence of Environmental Hazards*, Boulder, CO: Westview Press.

Moore, T. 1988. Planning Without Preliminaries, *Journal of the American Planning Association* 54(4), 525–8.

Morris, J. 1994. Involving Local People in Urban Renewal. *Town and County Planning* 62(9), 246–9.

Muller, J. 1987. The Concept of Error in Cartography, *Catographics* 24, 1–15.

Murck, B. 1996. *Dangerous Earth: An Introduction to Geologic Hazards*, New York: Wiley and Sons.

Naess, A. 1973. The Shallow and the Deep: Long Range Ecology Movement Summary, *Inquiry* 16, 95–100.

Naess, P. 1993. Can Urban Development Be Made Enviornmentally Sound?, *Journal of Environmental Planning and Management* 36(3), 309–33.

Newman, P. and Kenworthy, J. 1989. *Cities and Automobile Dependence*, Aldershot, UK: Gower.

Nolan, R. 1979. Managing the Crisis in Data Processing, *Harvard Business Review*, March–April.

Nuhfer, E., Proctor, R., and Moser, P. 1993. *A Citizen's Guide to Geologic Hazards*, Arvada, CO: American Institute of Professional Geologists.

Oladipo, E. 1985. A Comparative Performance Analysis of Three Meteorological Drought Indices, *Journal of Climatology* 5, 655–64.

O'Looney, J. 1997. *Beyond Maps: GIS and Decision Making in Local Government*, Washington, DC: International City Management Association.

Omi, P., Wensel, L., and Murphy, J. 1979. An Application of Multivariate Statistics to Land Use Planning, *Forest Science* 25, 399–414.

Ortolano, L. 1984. *Environmental Planning and Decision Making*, New York: Wiley and Sons.

Owens, S. 1986. *Energy Planning and Urban Form*, London: Pion Ltd.

Palmer, W. 1965. Meteorological Drought. Research Paper no. 45, US Dept. of Commerce Weather Bureau, Washington, DC.

Park, C. 1980. *Ecology and Environmental Management*, Boulder, CO: Westview Press.

Peck, S. 1998. *Planning for Biodiversity*, Washington, DC: Island Press.

Pepper, D. 1984. *The Roots of Modern Environmentalism*, London: Croom Helm.

Percesepe, G. 1995. *Introduction to Ethics*, Englewood Cliffs, NJ: Prentice-Hall.

Petak, W. and Atkisson, A. 1982. *Natural Hazard Risk Assessment and Public Policy*, New York: Springer-Verlag.

Pielke, R. 1990. *The Hurricane*, New York: Routledge.

Pojman, L. 1990. *Ethics: Discovering Right and Wrong*, Belmont, CA: Wadsworth.

Principles in EIA Studies. *The Environmental Professional* 15(1), 125–38.

Quayle, R. and Doehring, F. 1981. Heat Stress: A Comparison of Four Different Indices, *Weatherwise* 34, 120–4.

Rachels, J. 1998. *Ethical Theory*, Oxford: Oxford University Press.

Raper, J. 1998. Georeferenced Four-Dimensional Virtual Environment Computers, Environment and Urban Systems 22(6), 529–39.

Rees, W. 1996. Revisiting Carrying Capacity: Area-based Indicators of Sustainability, *Population and Environment* 17(3), 195–215.

Rees, W. and Wackernagel, M. 1996. Urban Ecological Footprints: Why Cities Cannot Be Sustainable and Why They Are a Key to Sustainability, *Environmental Impact Assessment Review* 16, 223–48.

Regan, T. 1995. The Nature and Possibility of an Environmental Ethic, in *Introduction to Ethics*, ed. G. Percesepe, Englewood Cliffs, NJ: Prentice-Hall.

Rescher, N. 1998. *Predicting the Future: An Introduction to the Theory of Forecasting*, Albany, NY: State University of New York Press.

Richardson, G. and Pugh, A. 1983. *Introduction to Systems Dynamics Modeling with Dynamo*, Cambridge, MA: MIT Press.

Ritter, D. 1986. *Process Geomorphology*, Dubuque, IA: William C. Brown Publishers.

Roberts, N. et al. 1983. *Introduction to Computer Simulation*, Reading, MA: Addison Wesley.

Robinson, J. 1988. Unlearning and Backcasting: Rethinking Some of the Questions We Ask about the Future, *Technological Forecasting and Social Change* 33, 325–38.

Rolston, H. 1988. *Environmental Ethics: Duties to and Values in the Natural World*, Philadelphia: Temple University Press.

Ross, B. 1989. Scenarios for Repository Safety Analysis, *Engineering Geology* 26, 285–99.

Saaty, T. 1977. A Scaling Method for Priorities in Hierarchical Structures, *Journal of Mathematical Psychology* 15, 234–81.

Sargent, J. 1971. Towards a Dynamic Model of Urban Morphology, *Economic Geography* 48, 357–74.

Sawicki, D. 1988. Policy Analysis, in Cantanese, A. and Snyder, J. (eds.), *Urban Planning*, New York: McGraw-Hill.

Schwartz, P. (1991). *Art of the Long View*, New York: Doubleday.

Shafer, T. 1977. *Urban Growth and Economics*, Reston, VA: Reston Publishing.

Shannon, R. 1975. *Systems Simulation: The Art and Science*, Englewood Cliffs, NJ: Prentice-Hall.

Shiffer, M. 1995. Environmental Review with Hypermedia Systems, *Environment and Planning B: Planning and Design* 22, 359–72.

Smith, G. 1989. Defining Managerial Problems: A Framework for Prescriptive Theorizing. *Management Science* 35(8), 963–81.

Smith, K. 1992. *Environmental Hazards*, London: Routledge.

Smith, P. 1995. A Fuzzy Logic Evaluation Method for Environmental Assessment, *Journal of Environmental Systems* 24(3), 275–98.

Smith, T. 1987. Requirements and Principles for the Implementation and Construction of Large-Scale GIS, *International Journal of Geographical Information Systems* 1, 13–31.

Soule, P. 1992. Spatial Patterns of Drought Frequency and Duration in the Contiguous USA, *International Journal of Climatology* 12, 11–24.

Spaling H. and Smit, B. 1993. Cumulative Environmental Change: Conceptual Framework, Evaluation Approaches, and Institutional Perspectives, *Environmental Management* 17(5), 587–600.

Spriet, J. and Vansteenkiste, G. 1982. *Computer-aided Modelling and Simulation*, London: Academic Press.

Stafford, S. 1993. Integration of Scientific Information Management and Environmental Research, in *Environmental Information Management and Analysis*, ed. W. Michner, London: Taylor and Francis.

Steiner, F. 1991. *The Living Landscape: An Ecological Approach to Landscape Planning*, New York: McGraw-Hill.

Steiner, F., Pease, J., and Coughlin, R. 1994. *A Decade with LESA: The Evolution of Land Evaluation and Site Assessment*, Kansas City, MO: Soil and Water Conservation Society.

Steinitz, C. 1977. Hand-drawn Overlays: Their History and Prospective Use, *Landscape Architecture* 66, 444–55.

Stern, P. and Fineberg, H. (eds.) 1996. *Understanding Risk*, Washington, DC: National Academy Press.

Stivers, C. 1990. The Public Agency as Polis, *Administration and Society* 22(1), 86–105.

Stokes, S. 1989. *Saving America's Countryside*, Baltimore, MD: Johns Hopkins University Press.

Storie, R. 1978. *Storie Index Soil Rating*, Special Publication Division of Agricultural Science, University of California, no. 3202.

Sukopp, H., Numata, M., and Huber, A. 1995. *Urban Ecology as the Basis of Urban Planning*, The Hague, Netherlands: SPB Academic Publisher.

Suter, G. et al. 1987. Treatment of Risk, *Environmental Impact* 11, 295–303.

Swartzman, D., Linoff, R., and Croke, K. 1982. *Cost Benefit Analysis and Environmental Regulations*, Washington, DC: The Conservation Foundation.

Talen, E. 1996. Do Plans Get Implemented? *Journal of Planning Literature* 10(3), 248–59.

Thompson, M. 1990. Determining Impact Significance in Environmental Impact Assessment, *Journal of Environmental Management* 30, 235–50.

Tobin, G. and Montz, B. 1997. *Natural Hazards, Explanation and Integration*, New York: Guilford Press.

Toffoli, T. 1987. *Cellular Automata Machines: A New Environment for Modeling*, Cambridge, MA: MIT Press.

Tomlin, C. 1990. *Geographic Information Systems and Cartographic Modeling*, Englewood Cliffs, NJ: Prentice-Hall.

Turner, M. 1989. Landscape Ecology: The Effects of Pattern on Process, *Annual Review of Ecology and Systematics* 20, 171–97.

US Environmental Protection Agency. 1992. Toxics in the Community: National and Local Perspectives. Office of Pesticides and Toxic Substances (TS-779), EPA 560/4-91-014. Washington, DC: GPO.

Van Valey, T. and Petersen, J. 1987. Public Service Centers: The Michigan Experience, in J. Desario (ed.), *Citizen Participation in Public Decision Making*, ed. J. Desario, Westport, CT: Greenwood Press.

Veregin, H. 1989. Error Modeling for Map Overlay, in *Accuracy of Spatial Databases*, ed. M. Goodchild, London: Taylor and Francis.

Vlek, C. and Wagenaar, W. 1979. Judgment and Decision Making Under Uncertainty, in *Handbook of Psychonomics*, ed. J. Michon, Amsterdam: North-Holland.

Vlek, C., Timmerman, D., and Otten, W. 1993. The Idea of Decision Support, in *Computer-Aided Decision Analysis*, ed. S. Nagel, Westport, CN: Quorum Books.

Von Neumann, J. 1966. *Theory of Self-reproducing Automata*, Urbana, IL: University of Illinois Press.

Wack, P. 1984. *Scenarios: The Gentle Art of Reperceiving*, Cambridge, MA: MIT Press.

Wackernagel, M. and Rees, W. 1995. *Our Ecological Footprint: Reducing Human Impact on the Earth*, Philadelphia, PA: New Society Publisher.

Wang, F. 1994. The Use of Artificial Neural Networks in a Geographic Information System for Agricultural Land Suitability Assessment, *Environmental and Planning A* 26, 265–84.

Way, D. 1978. *Terrain Analysis*, New York: McGraw-Hill.

Webster, C. 1993. GIS and the Scientific Inputs to Urban Planning, *Environment and Planning B: Planning and Design* 20, 709–28.

Weinberg, A. S. 1998. The Environmental Justice Debate: New Agendas for a Third Generation of Research, *Society and Natural Resources* 6, 605–14.

Westman, W. 1985. *Ecology, Impact Assessment and Environmental Planning*, New York: Wiley and Sons.

White, R. and Engelen, G. 1997. Cellular Automata as the Basis of Integrated Dynamic Regional Modeling, *Environment and Planning B: Planning and Design* 24, 235–46.

Whyte, A. and Burton, I. 1980. *Environmental Risk Assessment, Scope 15*, New York: Wiley and Sons.

Wilson, I. 1978. Scenarios, in *Handbook of Futures Research*, ed. J. Fowles, Westport, CT: Greenwood Press.

Wright, J. et al. 1993. *Expert Systems in Environmental Planning*, New York: Springer-Verlag.

Zechhaser, R. and Viscuss, W. 1990. Risk with Reason, *Science* 248, 559–64.

Zhu, X., Healey, R., and Aspinall, R. 1998. A Knowledge-Based Systems Approach to Design of Spatial Decision Support Systems for Environmental Management, *Environmental Management* 22(1), 35–48.

Index